Fiber Optic Sensors

Critical Reviews of Optical Science and Technology

Critical Reviews
of Optical Science
and Technology

Volume CR44

Fiber Optic Sensors

Eric Udd
Editor

Proceedings
of a conference held
8–11 September 1992
Boston, Massachusetts

Sponsored by
SPIE—The International Society for Optical Engineering

SPIE OPTICAL ENGINEERING PRESS

A Publication of SPIE—The International Society for Optical Engineering
Bellingham, Washington USA

Library of Congress Cataloging-in-Publication Data

Fiber optic sensors : proceedings of a conference held 8–11 September 1992,
  Boston, Massachusetts / Eric Udd, editor ; sponsored by SPIE—The International
Society for Optical Engineering.
        p.   cm. — (Critical reviews of optical science and technology ; v. CR44)
      ISBN 0-8194-0979-0 (hardcover) — ISBN 0-8194-0980-4 (softcover)
      1. Fiber optics—Congresses. 2. Optical detectors—Congresses.
    I. Udd, Eric. II. Society of Photo-Optical Instrumentation Engineers.
    III. Series
    TA1800.F51345 1992
    681'.2—dc20   92-35153
                                                                                        CIP

Published by
SPIE—The International Society for Optical Engineering
P.O. Box 10, Bellingham, Washington 98227-0010 USA
Telephone 206/676-3290 (Pacific Time) • Fax 206/647-1445

Printed in the United States of America.

Cover illustration: *Diagram of a fiber Bragg grating sensor system, from the paper
"Multiplexed fiber optic sensors" by A. D. Kersey, p. 202.*

Contents

APPLICATIONS OF FIBER OPTIC SENSORS

Conference Committee

Conference Chair

Eric Udd, McDonnell Douglas Electronic Systems Company

Session Chairs

Session 1—Components
Eric Udd, McDonnell Douglas Electronic Systems Company

Session 2—Discrete Fiber Optic Sensors
Gordon L. Mitchell, MetriCor, Inc.

Session 3—Interferometric Fiber Optic Sensors
John W. Berthold III, Babcock & Wilcox Company

Session 4—Distributed and Multiplexed Fiber Sensors
Juichi Noda, NTT International Corporation (Japan)

Session 5—Applications of Fiber Optic Sensors
William B. Spillman, Jr. University of Vermont

Preface

Over the past decade fiber optics technology has revolutionized the telecommunications market and is rapidly becoming a major player in such areas as cable TV and local-area networks. A similar revolution has taken place in the optoelectronics industry, with light-emitting and laser diodes playing key roles in such items as compact disc players, laser printers, pointers, and recently scanners.

Fiber optic sensor technology has developed in parallel with these industries and has benefited greatly from the availability of low-cost, high-performance components associated with mass production. This synergy has resulted in the introduction of fiber optic sensor products in the form of fiber optic gyros, biomedical sensors, and fiber sensors for process control. As the quality and quantity of low-cost, high-performance components continue to increase, the number of applications where fiber optic sensor technology can be effectively applied will continue to unfold.

This critical review contains a series of review papers intended to provide an overview of the state of the art in fiber optic sensor technology and the application areas that are beginning to emerge.

Eric Udd
McDonnell Douglas Electronic Systems Company

SESSION 1

Components

Chair
Eric Udd
McDonnell Douglas Electronic Systems Company

Passive Components for Fiber Optic Sensors

V. J. Tekippe

Gould Inc., Fiber Optics Division
6740 Baymeadow Drive, Glen Burnie, MD 21060 USA

ABSTRACT

Ironically, many of the passive components used extensively in telecommunications systems today were developed originally for fiber optic sensor systems. While the commercial sensor applications have developed more slowly, the breadth and scope of such applications continue to push the design engineer to develop new components. These developments have included, for example, splitters using hard clad silica fiber for medical and environmental sensors, improved splitters using polarization preserving fiber for gyroscopes, and unique wavelength division multiplexers for medical sensors. This paper will review the state of the art of passive fiber optic components with particular emphasis on sensor applications.

1. INTRODUCTION

The subject of fiber optic sensors has a rich history and an equally rich literature. Much of the early work, going back over a decade, was stimulated by military programs such as the Navy's Fiber Optic Sensor System (FOSS) program. The extensive review article published by Giallorerizi et al[1] in 1982 remains even today one of the most complete expositions of fiber optic sensor principles, implementations, and applications. While the military applications have continued to drive fiber optic sensor system development, notably in the areas of fiber optic gyroscopes and sonar systems, significant progress has been made in medical and industrial sensor applications[2] over the last ten years.

The early development of fiber optic sensors created a need for specialized fiber optic components to implement such sensors. Interferometric sensors for example, required singlemode splitters and couplers[3], two color intensity sensors required multimode wavelength division multiplexers[4], and fiber optic gyroscopes required phase modulators[5] as well as components made with polarization maintaining fiber.[6] While many of these components were initially developed for sensor applications, the advent of telecommunication and local area network applications of fiber optics during the middle eighties drove the widespread commercialization and mass production of many of these same components. Although the latter applications continue to push the development of fiber optic components, it is the fiber optic sensor system requirements that continue to provide the impetus to extend the capabilities of such components and account for much of the research and development in this area.

Perhaps the best way to explore the present day requirements for fiber optic sensor components is to examine a number of representative systems. However, before this is undertaken, a few general comments about fiber optic sensor systems are in order. By definition, a fiber optic sensor system uses light to probe an external perturbation and utilizes optical fibers to transmit the light. The most general configuration of such a sensor consists of a light source, a sensor system, and an optical detector, interconnected with optical fiber, as depicted in Figure 1a. The sensor system itself can either be intrinsic or extrinsic with respect to the optical fiber. In an intrinsic sensor, the fiber itself is acted upon by the external perturbation to ultimately cause intensity fluctuations in the light being transmitted by the fiber, as shown for example, by the microbend sensor in Figure 1a. In an extrinsic fiber optic sensor, the optical fiber is used to transmit the light to and from the sensor but the perturbation changes the intensity of the light external to the fibers themselves. The moving ball lens between two optical fibers, shown in Figure 1a, is an example of such an extrinsic sensor.

Since light provides the means to probe the external perturbation in a fiber optic sensor, the various properties of a light wave determine the type of sensor that can be utilized. Figure 1b shows the equation of the electric field of an electromagnetic wave and illustrates the properties of it that can be used to probe an external perturbation. The simplest type of sensor system is one

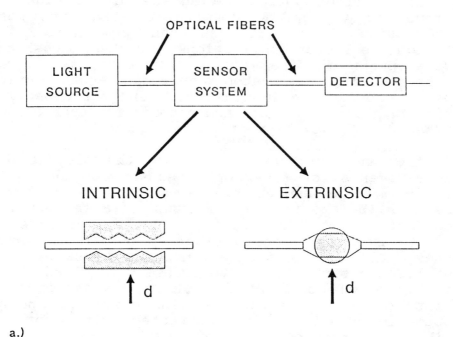

a.)

b.)

Figure 1 a) Generalized fiber optic sensor dia-
    gram. b) Electric field of a light
    wave and the parameters used for
    sensing mechanisms.

that directly changes the amplitude of the light wave resulting in an intensity fluctuation at the square law detector. Both of the sensor systems illustrated in Figure 1a, for example, are of this type. One of the greatest challenges in the design of this type of sensor is to separate the intensity fluctuations due to the external perturbation from the intensity fluctuations due to other causes. The interferometric class of fiber optic sensors is based on the relative phase difference between two light waves. In general, the external perturbation causes a phase shift between the two light waves resulting in an intensity change when the waves are recombined. Inherent to the design of fiber optic interferometers is the ability to coherently split and recombine an optical wave. The frequency or wavelength of a light wave can play a role in both intensity sensors, because of wavelength selective absorption or reflection, and in interferometric sensors since the phase shift is wavelength dependent. Finally, the vectorial nature of a light wave and its resultant polarization properties can also be used to induce intensity changes in a sensor system. It is interesting to note that since, in most cases, a square low detector is used to measure the effects of the perturbation, it is necessary to design a sensor system so that the external perturbation ultimately causes a change in intensity of the light, independent of the parameter of the light wave that is influenced by the perturbation. The rich variety of innovative and creative fiber optic sensor systems which have been reported are all a result of using these various parameters of a light wave, either singly or in combination, to achieve this result.

Figure 2 shows a schematic diagram of a rotary position sensor for airborne applications and is an example of an extrinsic fiber optic intensity sensor.[7] This sensor uses a multimode fiber optic splitter/combiner to separate the incoming light wave into signal and reference beams. The position of the rotating code disk is determined by the intensity of the reflected signal which is ratioed to the constant reflection of an outer ring of the code disk. This ratio method eliminates the effects due to fiber and connector losses in the system. In this case the splitter/combiner had to be quite small and operate at temperatures up to 300°C in an aircraft engine environment.

An example of a two color intensity fiber optic sensor system is illustrated in Figure 3.[8] While two different colors or wavelengths are often used to provide signal and reference beams, in this implementation the two colors are used to access two separate intrinsic fiber optic sensors via single input and output fibers. Part of the input

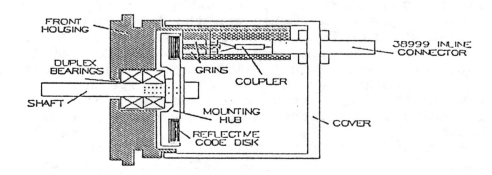

Figure 2. Schematic diagram of a fiber optic rotatory position sensor.

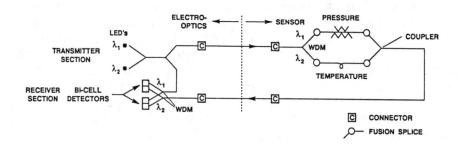

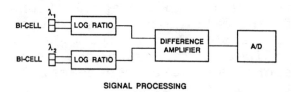

Figure 3. Diagram of a dual fiber optic sensor system using wavelength division multiplexing and single fibers to and from the sensor.

beam is used as a reference beam to eliminate source variations. Successful implementation of this system clearly depends on the availability of stable multimode splitters, combiners and wavelength division multiplexers (WDM).

Figure 4 shows an example of an intrinsic, distributed single mode fiber optic sensor which uses

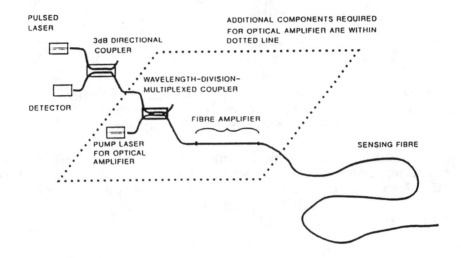

Figure 4. Distributed fiber optic sensor using OTDR techniques in conjunction with a fiber amplifier.

optical time domain reflectometer (OTDR) techniques for sensing perturbations to the sensing fiber.[9] The range of this sensor is significantly enhanced by the use of an in-line optical amplifier to boost both the outgoing and return signals. In this system a singlemode 3 dB directional coupler is used to separate the return light from the incoming light and a singlemode WDM is used to inject the pump light into the fiber amplifier.

The Mach-Zehnder fiber optic interferometer, shown schematically in Figure 5, is probably the most common example of an intrinsic fiber optic interferometric sensor.[10] It has been used for a variety of sensor applications to measure, for example, temperature,[11] pressure,[12] displacement,[13] and magnetic fields,[14] among others. This sensor

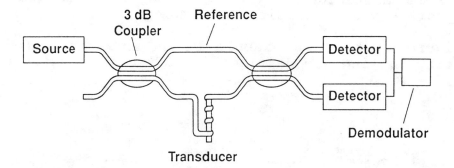

Figure 5. Schematic diagram showing the
configuration of a typical
Mach-Zehnder fiber optic sensor.

system uses 3 dB singlemode couplers to split the
incoming beam into a sensor and a reference fiber
and then to recombine the light waves after the
transducer has introduced a phase shift into the
light in the sensor fiber.

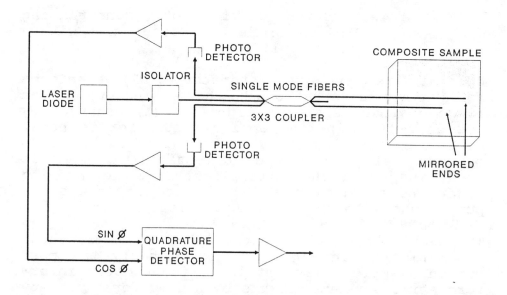

Figure 6. Michaelson interferometer fiber optic
sensor which ultilizes a 3x3 coupler.

Figure 6[15] illustrates the fiber optic implementation of a Michaelson interferometer for use in "smart skins". The two sensor fibers are embedded in the composite material and hence the difference in path length is sensitive to the mechanical properties of the material. The 3x3 singlemode splitter not only provides a means to get the light into the sensing fibers but also provides two outputs for the quadrature phase detection scheme.

The schematic representative of a fiber optic Sagnac interferometer used for a fiber optic gyro (FOG) application is shown in Figure 7.[16] The counter rotating optical beams in the polarization maintaining fiber coil undergo a relative phase shift if the coil itself is rotating in the plane of the coil. While a 3 dB singlemode fiber optic coupler (possibly made from polarization maintaining fiber) is used to get light into and out of the gyro, the integrated optic chip, which provides for splitting the light into two beams and the introdution of separate phase shifts into these beams, is the real heart of this system.

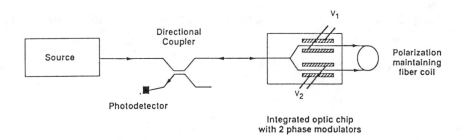

Figure 7. Diagram of a fiber optic rotation sensor using an integrated optic FOG chip.

It is clear from the representative examples given above that, aside from conventional optical components, the majority of the requirements for the components necessary to implement fiber optic sensor systems fall into the categories of single-mode and multimode splitters, combiners, and wavelength division multiplexers as well as integrated optic modulators. The remainder of this paper will be devoted to these components with an aim toward delineating the current state of the art as well as some of the recent developments for sensor system applications.

2. SPLITTERS/COMBINERS

2.1 Multimode Splitter/Combiner

As the examples in the previous section showed, many of the multimode intensity sensors depend on splitters and combiners to get light into and out of the sensor system and/or to provide a reference beam to eliminate spurious intensity effects. These splitters/combiners are commercially available using bulk component technology, fused biconical taper technology, and integrated optic technology. Each technology has its own advantages and disadvantages.

The oldest technology for fabricating multimode splitters and combiners is the bulk optic technology. As the name implies, these components usually require the light to be removed from the fiber so that the splitting can be achieved by a conventional optical component with the split or combined light then reinserted into optical fiber(s). One example of such a component is shown in Figure 8.[17] In this SELFOC lens coupler, SELFOC lenses and a half mirror are used to split the incident beam into transmitted and reflected beams. The input and

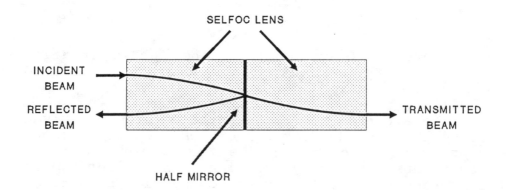

Figure 8. SELFOC lens fiber optic splitter.

output fibers are usually attached to the SELFOC lenses using an adhesive. The mirror can be made on the end of one of the SELFOC lenses before the two lenses are attached. Figure 9 shows a more nearly all-fiber construction of a multimode

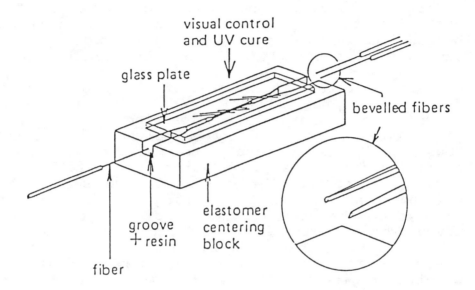

Figure 9. All fiber construction of a butt-joined multimode fiber optic splitter.

splitter.[18] In this case the two output fibers are polished halfway through and then butt-joined to the input fiber using an index matching adhesive. An elastomer block is used during the assembly process and a glass plate is attached to the fibers to provide stability in the final device. The elastomer block is removed after the adhesive is cured.

The advantages of these components are that they can be used with a variety of different fiber types, are modally insensitive, and are relatively inexpensive. The chief disadvantages are that they tend to have somewhat higher insertion losses, are somewhat large, and because the critical fiber alignments are maintained by adhesives, are more environmentally sensitive. In particular, with regard to the latter point, these components would not be suitable for high temperature applications.

Multimode splitters have also been made by the fused biconical taper technique for quite some-time.[19] These components generally have greater environmental stability than bulk components and are wavelength independent, but, because of the nature of the coupling process, tend to be more modally sensitive. Good modal insensitivity for 50/50 splitters can only be achieved when special precautions are taken to extend the coupling process until essentially all the modes participate.

Of recent interest in the area of multimode sensors for medical and industrial applications is the introduction of hard clad silica fiber in which the glass core is cladded by a polymeric coating. However, it is difficult to make satisfactory splitters from this fiber using bulk optic tech-niques because the adhesives used in this approach will not bond well to the cladding and will extract light from the core if the cladding layer is re-moved. The most successful technique used with this fiber has been the fused biconical taper techni-que.[20] In this case the cladding is removed in the coupling region and the two glass cores are fused together giving a device with a very stable cou-pling ratio. Figure 10 shows the excellent stabili-ty of the coupling ratio (dotted line) over a temperature range from -30°C to +125°C for a hard clad silica fiber fused coupler.

Modal sensitivity, alluded to earlier, is of funda-mental concern to multimode fiber optic sensor system designers because quite often these modal effects are observed as spurious intensity changes. Aside from the modal sensitivity of the components themselves, the system designer must also be aware of modal effects arising both from the fiber and from the launch conditions. Hard clad silica fiber, for example, since it is coated with a polymer whose index of refraction has a much greater change in index with temperature then the core glass, will exhibit a variation in numerical aperture (NA) with temperature. Thus a fiber which experiences fully filled launch conditions at one temperature will be underfilled or overfilled at another temperature. Cladding modes which are developed in the over-filled condition will usually be stripped out by

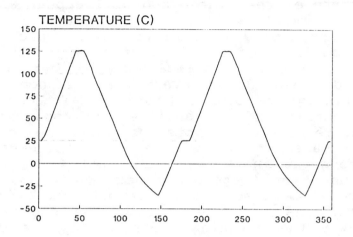

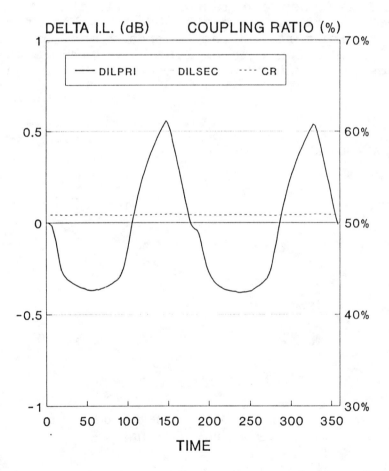

Figure 10. Temperature dependence of the coupling ratio and insertion loss of a fused coupler made with hard clad silica fibers.

other parts of the system and will appear as loss to the system. This is illustrated quite dramatically by the solid line in Figure 10 which shows the change in insertion loss with temperature for the same hard clad silica fused fiber coupler discussed above. In an over-filled launch condition, although the coupling ratio remains constant, the system will exhibit an insertion loss which is temperature dependent. Note that when the temperature increases the NA of the fiber increases and the throughput increases while just the opposite occurs when the temperature decreases. Modal effects can also arise from the launch condition into the fiber. As discussed above, the coupler can strip away the cladding modes making the device look lossy when the launch exceeds the NA of the fiber. Figure 11 shows the insertion loss for both the primary port and the secondary port of a fused coupler as a function of launch NA.[20] While the two curves are together over a considerable range of

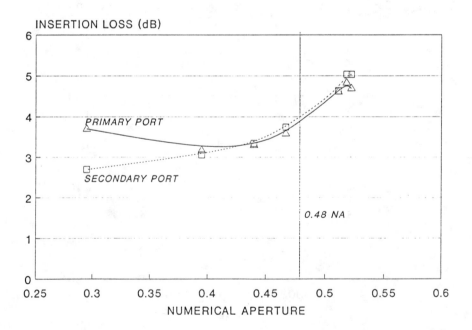

Figure 11.   Insertion loss as a function of NA for the two output ports of a fused coupler made with hard clad silica fiber.

NA, indicating a stable coupling ratio, the insertion loss clearly increases when the launch exceeds the NA of the fiber. This same figure also illustrates another effect of launch NA. The divergence of the two curves when the fiber is underfilled shows that the coupling ratio is sensitive to the launch conditions, particularly if the launch is off axis. Figure 12 clearly illustrates this effect. When the fiber was fully filled using an on-axis launch, the two output fibers show uniform modal distribution patterns (solid lines). When the launch was moved off-axis, the output of the primary fiber is dominated by the lower order modes and the output of the secondary fiber is dominated by the higher order modes.

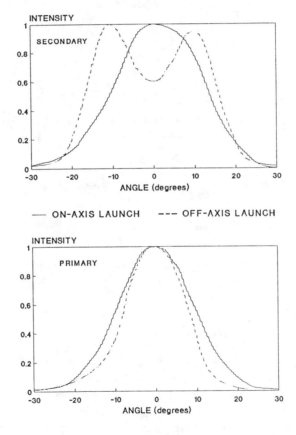

Figure 12.   Far-field modal distribution of the output ports of a fused multimode coupler for on-axis launch conditions (solid lines) and off-axis launch conditions (dotted lines).

In many fiber optic sensor systems, space is at a premium and yet the components must operate over large ranges of environmental conditions. As mentioned earlier, the rotation sensor of Figure 2 required a multimode splitter which was small enough to fit within the sensor but would operate up to 300°C. The coupler developed for this application is shown in Figure 13.[21] It is a fused coupler made from high temperature polyimide coated fiber with a total packaged length of about 1 cm. Despite its small size, it operated successfully over the full temperature range of 0°C to 300°C.

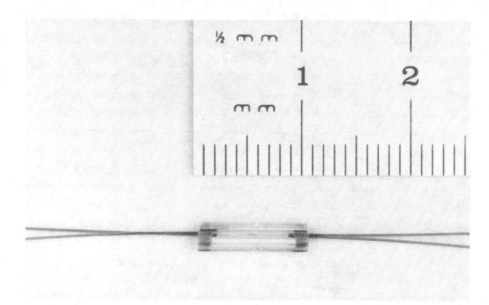

Figure 13. Photograph showing a minature fused multimode coupler for use up to 300°C.

Another area which is gaining increased interest is the use of multimode fiber optic sensors made from plastic optical fibers.[22] While the loss of plastic optical fibers is still relatively large compared to telecommunication grade fiber, it is being very favorably considered for short distance sensor applications, for example in automobiles. As with any other type of fiber, successful implementation of plastic optical fibers for sensor applications depends on the development of components made from this fiber and considerable work has been done in this area in recent years. Figure 14 for example, shows an optical waveguide structure molded from

PMMA with access ports into which plastic optical fibers are inserted.[23]

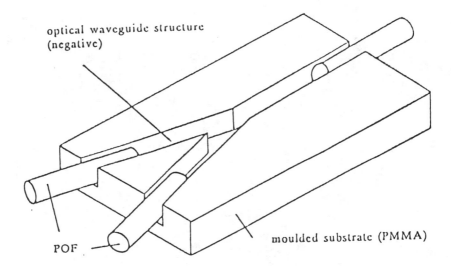

optical waveguide structure
(negative)

POF

moulded substrate (PMMA)

Figure 14. Optical coupler for plastic optical fiber which uses an optical waveguide structure molded from PMMA.

2.2 Singlemode splitters/combiners

While bulk component techniques have been applied to singlemode fibers, the extreme dimensional control required is difficult to achieve and maintain and hence the commercialized singlemode components have been dominated by the fused biconical taper technique. These latter components are readily available from a number of suppliers and, with the advent of large scale production for the telecommunication market, have reached a high level of sophistication at modest prices. Recent advances in singlemode component technology for sensor applications include high temperature couplers, miniature couplers, and fused couplers for polarization maintaining fibers. In addition, integrated optic and semiconductor processing techniques have been combined to produce singlemode components.

Figure 15 shows the coupling ratio and insertion loss for a fused singlemode coupler over the temperature range from 0°C to 200°C. This coupler, made with polyimide coated fiber, not only shows excellent stability for the coupling ratio but

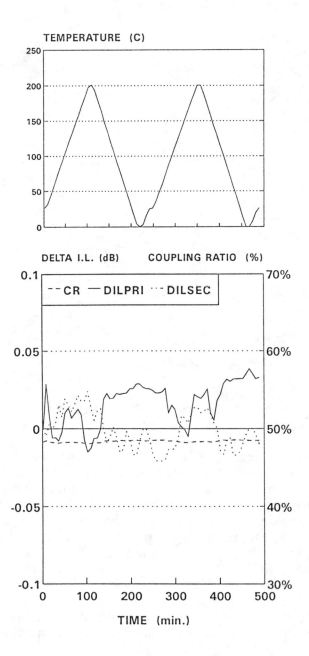

Figure 15.   Temperature dependence of the coupler ratio and the change in insertion loss of a fused singlemode coupler between 0°C and 200°C.

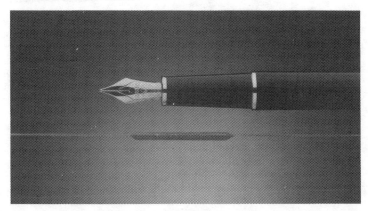

Figure 16.   Photograph of a miniature fused
             singlemode coupler made with payout
             fiber.

exhibits changes in insertion loss of less then
0.05 dB over the full temperature range.

As with the multimode components discussed above,
size is also an important consideration for single-
mode components used in sensor applications. Figure
16 shows a fused singlemode coupler made out of
"payout fiber" which has a length of about 1 inch
and a diameter of about 1/8 inch. Payout fiber,
which was first introduced for tethered vehicle
applications, has superior bend characteristics and
a smaller size (80 μm OD). The latter characteristics allow for smaller fused components. Some fiber
optic gyroscopes, where space is a premium, require
splitters made from polarization maintaining fibers. A similarly small fused coupler with polarization maintaining fibers, as shown in Figure 17,
has also been recently reported.[24]

In the last few years the fused biconical taper
technique has pushed beyond fusing just two fibers
and today components are available commercially
with up to six output fibers. The 3x3 splitter is
particularly useful for sensor applications as
noted in regard to Figure 6. The cross-section of a
fused 3x3 coupler is shown in Figure 18.[25] This
coupler not only allows for light input along one
fiber to be split among the other two fibers but
allows for two output ports in the reflective
Michaelson interferometric configuration.

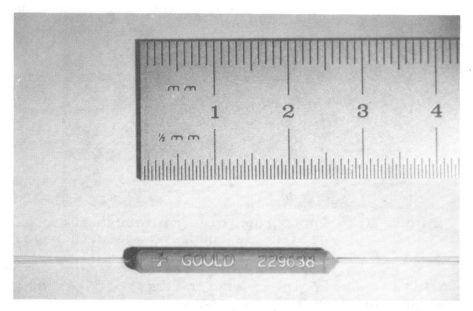

Figure 17. Photograph of a miniature singlemode coupler made with polarization maintaining fibers.

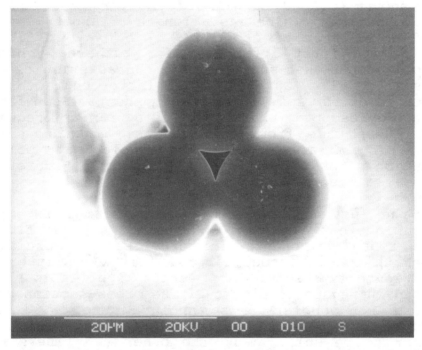

Figure 18. Cross section of a fused 3x3 single-mode coupler.

Higher output port splitters are important for a number of distributed sensors where a single source is used to power a large number of parallel sensors. While practical fused devices are currently limited to six output ports, devices using integrated optic techniques have been reported with up to sixteen output ports.[26] Two different approaches have been used to produce commercial devices, namely, the diffusion of dopants into glass substrates and silica on silicon waveguides. Figure 19 illustrates the process used by Corning in which buried

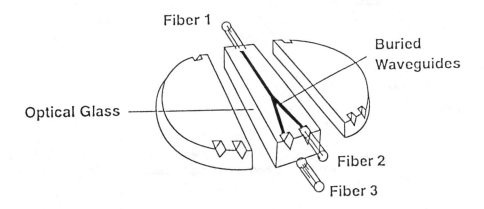

Figure 19.   Schematic diagram of a integrated optic coupler made with diffused waveguides in a glass substrate.

waveguides in glass are obtained by diffusion through a photolithographic mask in a molten salt solution. Once the waveguides are created, optical fibers are aligned with the waveguide and attached with adhesvies to the substrate. While the device exhibits good insertion loss and uniformity, and is fairly independent of wavelength and polarization, the fiber attachment is difficult to achieve and maintain over a wide range of environmental conditions.

3. WAVELENGTH DIVISION MULTIPLEXERS

3.1 Filter based components

Fused multimode couplers are wavelength independent and hence most multimode WDM's are of the bulk component type and use conventional optical filter technology to achieve WDM performance. Figure 20

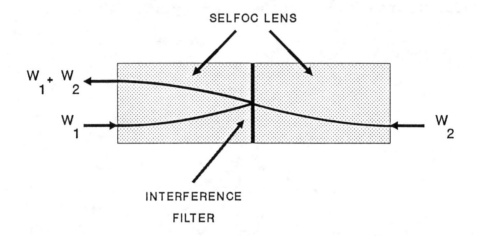

Figure 20. SELFOC lens wavelength division multiplexer which uses an interferometric dichroic filter.

shows a schematic diagram of a multimode WDM multiplexer which uses SELFOC lenses and an interference filter.[17] The interference filter is transmissive at wavelength W_2 and reflective at wavelength W_1. Thus input fibers at W_1 and W_2 will result in both wavelengths being present in the output fiber. The same component can be used as a demultiplexer by reversing the input and the outputs. An all fiber implementation of a multimode WDM is shown in Figure 21.[18] In this case the dichroic filter is placed directly on the ends of the fibers which are polished at an angle. This figure shows the demultiplexer configuration in which the input light containing wavelengths W_1 and W_2 is split according to wavelength in the two output fibers. This device is constructed in a similar manner to the coupler shown in Figure 9.

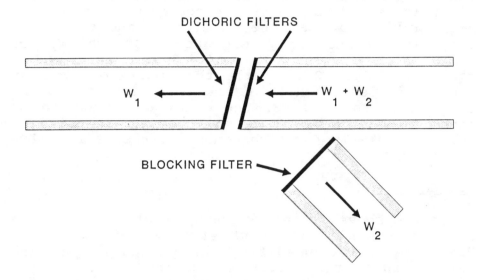

Figure 21. All fiber construction of a multi-mode wavelength demultiplexer which uses dichroic filters deposited directly on the fiber ends.

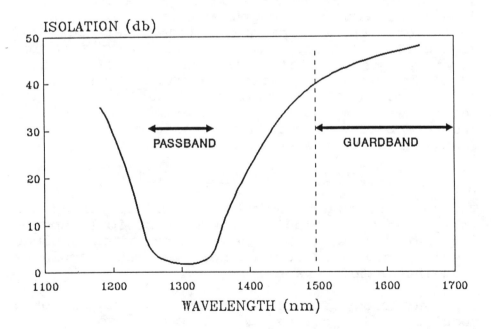

Figure 22. Filter characteristics of a dichroic filter used in bulk component wavelength division multiplexers.

The chief advantage of these devices is that the dichroic filter can be custom designed to give very high transmission at one wavelength and very high refection at another wavelength. A typical filter response curve for a 1300 nm/1550 nm filter is shown in Figure 22.[17] While the insertion loss at 1300 nm is less then 1 dB, the isolation at 1550 nm is greater then 40 dB. The chief disadvantage of these components are their environmental sensitivity, particularly for high temperature sensor applications.

3.2 Fused WDM

While the above filter technologies have also been applied to singlemode components, fused singlemode WDM's are usually used for this application and are well known in the telecommunication industry. The WDM characteristics in a fused WDM device arise from the enhancement of the wavelength dependent coupling of such components.[27] Figure 23 shows the coupling ratio as a function of wavelength for a fused WDM designed to inject the 1480 nm pump light into a 1550 nm optical amplifier as shown in Figure 4. The coupling ratio exhibits a nominal $\sin^2$ dependence with wavelength and this dictates the equivalent filter response. The lower part of Figure 23, which shows the isolation as a function of wavelength, exhibits a maximum value of about 25 dB of isolation in practical devices with a pass-band of about 40 nm at 18 dB of isolation. Thus, while these components exhibit excellent stability and versatility, their filter performance is somewhat limited by the coupling mechanism.

4. INTEGRATED OPTIC MODULATORS

The third class of components to be discussed are integrated optic modulators. While strictly speaking not a passive component, they often include passive components such as splitters and combiners when integrated on a chip.

The technology for fabricating integrated optic modulators by diffusing titanium into lithium niobate has been known for some time.[28] However, advancements are continually being made to reduce size and drive voltage requirements, improve the

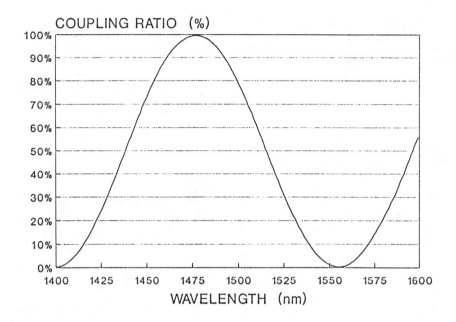

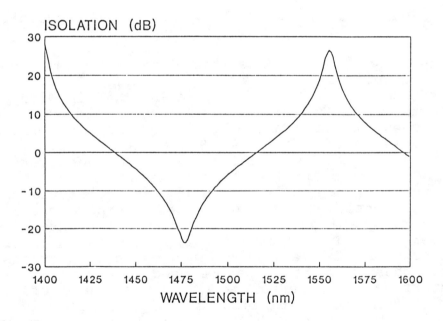

Figure 23.   Coupling ratio and isolation of a
             fused wavelength division multi-
             plexer designed for pumping optical
             amplifiers.

performance bandwidth, and to make them more compatible with fiber optic sensor applications by fiber pigtailing techniques. More recently, the waveguides have been fabricated using an annealed proton exchange technology in order to achieve enhanced extinction ratios of more then 60 db.[29]

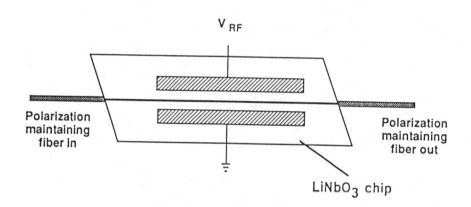

Figure 24. Integrated optic phase modulator chip.

Figure 24 shows a schematic diagram of an integrated optic phase modulator.[30] It consists of a Ti diffused waveguide region in a LiNbO$_3$ chip with an electrode on either side. The ends are cut and polished and have polarization maintaining fiber pigtails attached to them. The angled end significantly reduces back reflections from the crystal. An AC electric field applied between the electrodes causes a change in index in the waveguide via the electro optic effect resulting in a phase modulation of the light wave. Polarization maintaining fiber is used to define the input state of polarization to achieve well defined phase modulations. Such phase modulators can be used in one arm of an interferometric sensor to induce DC phase offset and/or an AC modulation signal to effectively achieve a frequency shift (serrodyne modulation.)[28]

Commercially available[30] phase modulators made using the annealed proton exchange process offer bandwidths up to 2GHz with half wave voltages of about 5V, and exhibit insertion losses less than 4.5 dB with polarization maintaining pigtails. Backreflection and polarization extinction on the chip is typically greater then 60 dB while the polarization crosstalk of the fiber itself is on the order of 30 dB. Such devices are still quite expensive in single quantities but are expected to have significantly reduced prices in volume.

In addition to simple phase modulations, integrated optic chips which contain more than one component are also available. For example, a dual output amplitude modulator is available which uses a Y-branch splitter, a phase modulator and a passive 3 dB coupler to achieve complimentary, amplitude modulated signals.[31] The fiber optic gyroscope shown in Figure 7 uses a commercially available "FOG" chip as shown in Figure 25. It consists of a Y-branch splitter and two independent phase modulators pigtailed with polarization maintaining fiber.[30]

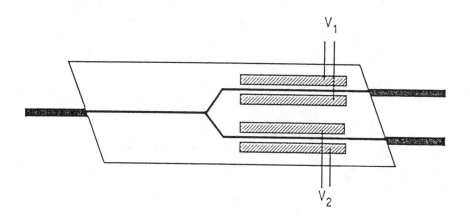

Figure 25. Integrated optic FOG chip for fiber optic gyroscope applications.

5. CONCLUSION

As the development of fiber optic sensors continues to push towards large scale implementation and production, so also has the need continued for fiber optic components with even higher performance and reliability. Much of the innovative research and development of fiber optic components is directed towards fiber optic sensor systems and such developments will continue to benefit all applications of fiber optics.

6. REFERENCES

1. Thomas G. Giallorenzi, Joseph A. Bucaro, Anthony Dandrige, G. H. Sigel, Jr., James H. Cole, Scott C. Rashleigh, and Richard G. Priest, "Optical Fiber Sensor Technology", IEEE J.Q.E., 18(4), 626-665, 1982.

2. Eric Udd, <u>Fiber Optic Sensors, An Introduction for Engineers and Scientists</u>, John Wiley & Sons, Inc., New York, 1991.

3. J. A. Bucaro, H. P. Dardy, and E. Carome, "Fiber optic hydrophone", J. Acoust. Soc. Amer., 62, 1302-1304, 1977.

4. T. Uchida and S. Sugimoto, "Micro-optic devices for optical communications", Proc. 4th ECOC (Geneva, Switzerland), 8-1, 1978.

5. K. Bohm, P. Martin, K. Petermann, E. Weidel, and R. Ulrich, "Low-drift fiber gyro using a superluminescent diode", Electron. Lett., 17, 352-353, 1981.

6. R. Ulrich, "Fiber-optic rotation sensing with low drift", Opt. Lett., 5, 173-175, 1980.

7. E. D. Park, W. J. Swafford, and B. K. Lamb, Proc. SPIE Conference on Fiber Optic Reliability: Benign and Adverse Environments, 1366, 294, 1990.

8. J. W. Berthold, III, "Status and review of fiber optic sensors in industry", <u>Fiber Optic Sensors</u>, ed. by Eric Udd, SPIE CR44-15, to be published.

9. J. P. Dakin, "Distributed fiber optic sensors", <u>Fiber Optic Sensors</u>, ed. by Eric Udd, SPIE CR44-10, to be published.

10. Peter Adrian, "Technical advances in fiber-optic sensors: theory and applications", Sensors, 8(9), 23-45, 1991.

11. S. J. Petuchowski, G. H. Sigel, and T. G. Giallorenzi, "Singlemode fiber point and

extended sensors", Electron. Lett. 18, 814, 1982.

12. J. A. Bucaro, N. Lagakos, J. H. Cole, and T. G. Giallorenzi, "Fiber optic acoustic transduction", <u>Physical Acoustics</u>, V. XVI ed. by W. P. Mason and R. N. Thurston, Academic Press, New York, 1982.

13. M. H. Slonecker, D. W. Stowe, V. J. Tekippe, and J. D. Beasley, "Effect of strain on light propagation in a singlemode optical fiber", Adv. in Ceramics, 2, 209, 1981.

14. A. Dandrige, A. B. Tveten, G. H. Sigel, Jr., E. J. West, and T. G. Giallorenzi, "Optical fiber magnetic field sensor", Electron. Lett., 16, 408-409, 1980.

15. Raymond M. Measures and Kexing Liu, "Fiber optic sensors focus on smart systems", IEEE Circuits & Devices, 8(4), 37-46, 1992.

16. F. J. Leonberger, T. K. Findakly and G. P. Suchoski, "LiNbO$_3$ and LiTaO$_3$ integrated optic components for fiber optic sensors", Proc. OFS '89, Paris, 1989.

17. Product literature, JDS Optics, Inc., PO Box 6706, Station J, Ottawa, Ont. Canada, K2A 3Z4.

18. Product literature, ATI Electronique, Zi de St. Guenault, 6 rue Jean Mermoz, 91080 Courcouronnes, France.

19. B. S. Kawasaki and K. O. Hill, "Low-loss access coupler for multimode optical fiber distribution networks", Appl. Opt., 16(7), 1794-1795, 1977.

20. P. A. Nagy and D. R. Moore, "Hard clad silica fused biconical taper couplers", Proc. of SPIE Conference on Fiber Optic Components and Reliability, ed. by Paul M. Kopera and Dilip K. Paul, 1580, 424-429, 1991.

21. P. A. Nagy, private communication.

22. Proceedings, Information Gatekeepers Conference on Plastic Optic Fiber and Applications, Paris, 1992.

23. A. Rogner, "Micromoulding of passive network components", Proc. Information Gatekeepers Conference on Plastic Optic Fibers and Applications, Paris, Pp. 102-104, 1992.

24. D. R. Campbell, Jr., "Minature fused polarization-maintaining couplers for gyroscope applications", Proc. SPIE Conference on Components for Fiber Optic Applications VI, 1792-14, to be published.

25. H. S. Daniel and D. R. Moore, "MxN Fused biconical tapered fiber couplers", Proc. SPIE Conference on Components for Fiber Optic Applications V, 1365, 53-59, 1990.

26. Product literature, Corning Incorporated, Opto-Electronic Components, MP-50-3-8, Corning, NY 14831.

27. V. J. Tekippe, "Passive Fiber Optic Components Made by the Fused Biconical Taper Process", Fifth National Symposium on Optical Fibers and Their Applications, Warsaw, SPIE 1805 (1989). Reprinted in Fiber and Integrated Optics, 9, 97-123, 1990.

28. Lonard M. Johnson, "Optical modulators for fiber optic sensors", <u>Fiber Optic Sensors, An Introduction for Engineers and Scientists</u>, ed. by Eric Udd, 99-137, John Wiley & Sons, Inc., New York, 1991.

29. T. Findakly and M. Branson, "High-performance integrated-optical chip for a broad range of fiber-optic gyro applications", Opt. Lett. 15(12), 673-675, 1990.

30. Product literature, United Technologies Photonics, Inc., 1289 Blue Hills Avenue, Bloomfield, CT 06002.

31. Product literature, New Focus, Inc., 340 Pioneer Way, Mountain View, CA 94041.

Optical fibres for sensors

J.E. Townsend and D.N. Payne

Optoelectronics Research Centre,
University of Southampton, SO9 5NH, UK

ABSTRACT

Optimisation of optical fibre design to fully realise the potential of optical sensing systems is discussed. In particular, waveguide geometry and host glass composition are discussed with reference to specific sensor applications.

1. INTRODUCTION

Silica based fibres have many features which make them potentially attractive for sensing applications including light weight, small size, strength, flexibility, chemically passive nature and insensitivity to electromagnetic interference. However optical fibre sensors are not cheap and will not be introduced unless they provide a more economic solution or performance not realisable by conventional methods.

Considerable scope exists for optimising the properties of silica fibres for sensor applications either by modifying the waveguide geometry or by altering the host composition. A few special fibres are currently commercially available of which the polarisation maintaining and rare-earth ion doped structures are the most widely known. The first is widely employed as an intrinsic sensor; the gyroscope, whilst the latter is most used as an active, laser medium and can be employed as a laser source for sensor systems.

The fibre current sensor is potentially attractive owing to its inherent safety in high field environments. Both host and geometric optimisation are discussed for maximum sensitivity and environmental stability. Distributed gratings in fibres are now commonly fabricated with high extinction ratios and narrow linewidths and have great potential as distributed sensors of, for example, temperature or strain. Rare-earth ion doped fibres offer potential as both passive or active distributed sensors but, to date, their greatest potential is as lasers offering application specific performance.

The pace of development is increasing as materials research is intensified and a number of fibre designs for specific applications have been developed. Waveguide geometry and host composition design criteria for optimised performance are described here for a selection of fibre sensors and laser sources to indicate the potential of fibre design.

2 WAVEGUIDE GEOMETRY

2.1 Fibre current sensor

The fibre current sensor based on a modified waveguide structure is presented as an example of the potential of novel geometries[1].

Fibre current sensors consist of a number of fibre coils wound around a current carrying conductor. By virtue of the Faraday effect the current flow induces a rotation of the plane of polarisation of the light travelling along the fibre. The rotation measured depends on the magnitude of the current, the number of fibre turns and the Verdet constant of the material whilst the measurement bandwidth is determined by the transit time of light through the coil. The problem central to the development of a practical device is that of linear birefringence in the fibre invariably present as a result of intrinsic manufacturing imperfections or bend and pressure induced birefringence. This birefringence dominates any Faraday induced rotation in conventional fibres. In a practical device current induced circular birefringence (rotation) should swamp packaging induced linear birefringence. This may be achieved by introducing intrinsic circular birefringence or by increasing the Faraday rotation[2], a form of circular birefringence to swamp any linear effects.

Both approaches have been tested and currently the circularly birefringent fibre performs best. The intrinsic loss of high Verdet constant fibres tends to be high (0.3dB/m) so compromising their sensitivity. A figure of merit for maximum current sensitivity, the ratio of Verdet constant to fibre loss has been developed[3] and it is clear that, for the optimum length silica remains superior. However, for high bandwidth applications the use of high Verdet constant glasses is potentially advantageous but, to date, packaging and form birefringence tend to limit performance although in principle, the use of zero-stress optics coefficient glasses may overcome this drawback.

The spun elliptically birefringent fibre[4] (SEB) is the most robust and practical of the designs considered[5-7] to date. SEB fibres are prepared by spinning a highly linearly birefringent fibre during the draw process to impart a rapid, built-in rotation of the fibre birefringent axis. By introducing a large net elliptical birefringence a compromise can be reached between polarisation holding (and hence packaging insensibility) and the electric current response of a true circularly birefringent fibre. These compromises, and practical fabrication problems, are discussed for production of an optimised fibre.

Expressions for elliptical beat length, L_p[1], and maximum current sensitivity, S, relative to an ideal, circularly birefringent fibre, can be derived[8] from the Jones matrices describing the net birefringence of the spun fibre and Faraday rotation per unit fibre length and are summarised graphically in Figure 1. The actual value of fibre sensitivity found in practice depends on the optical configuration used but here S represents the best case. The normalized elliptical beat length and relative current sensitivity for the fibre are plotted in Figure 1(a) and (b) as a function of the ratio of the spin pitch L_T to unspun linear beat length L_p. To make the fibre insensitive to external packaging effects the elliptical beat length L_p must be as short as possible. From Figure 1(a) it is clear that this is achieved by choosing a starting preform with short unspun linear beat length L_p and then selecting a large ratio of spin pitch to unspun linear beat length. However, this ratio cannot be increased indefinitely as the effect is a large decrease in the relative fibre current sensitivity (Figure 1(b)). The decreased sensitivity results because the ellipticity of the eigenmodes of the fibre becomes greater and approaches that of a high linearly-birefringent fibre (i.e., linearly polarized modes) which has a very small response to current. A compromise between the two parameters must be chosen to obtain as short an elliptical mode beat length as possible without seriously reducing the sensitivity. An optimized fibre might have a spin pitch approximately equal to the unspun linear beat length, and is quasicircularly birefringent with $L_p^1 \approx 4.2L_p$ with relative fibre current sensitivity S around 80 per cent (see points marked on Figure 1(a) and (b)). A typical unspun bow-tie preform has a fibre beat length of 1mm, so a spun fibre having an elliptical beat length of 4.2mm will result. This degree of elliptical birefringence is very high and sufficient to allow a bend radius as small as 5mm, making the fibre attractive for short length, high bandwidth applications.

To achieve a spin pitch of 1mm fibre must be drawn slowly whilst the preform is spun rapidly. In practice a spin twist rate of ~2,000rpm is employed with a fibre drawing speed of only a few metres per minute. At such low speeds considerable care is required to ensure the quality of uv cured acrylate coating is maintained and the core is not exposed to high levels of uv light which induce excess loss. The high spin rates require a perfectly straight precursor preform in order to minimise oscillations in the preform tip position which will adversely affect fibre yield. Further, accurate control of fibre drawing temperature and tension are essential. Nevertheless lengths of several hundred meters can be drawn with accurate control of diameter and pitch.

High performance devices with low temperature sensitivity (0.05%/ °C) have been demonstrated in a number of configurations. Typical performance for the SEB fibre current sensor is shown in Figure 2 where the current sensitivity is plotted as a function of bandwidth (or length). The curves are plotted for two realizable coil diameters of 10 and 100mm and assuming a practical fibre length limit of 100m, which corresponds to 3,200 turns of a 10mm diameter coil. It is clear that, in the low-frequency region (<300 kHz), the fibre length is limited to 100m by practical, not bandwidth considerations. The minimum detectable current is limited by detection shot noise and thus increases with (bandwidth)$^{1/2}$. Typically for a 1kHz measurement bandwidth a maximum sensitivity of 100μA rms is obtained for a 10mm coil. For higher frequencies (1-100MHz) the sensor is still detection shot-noise limited but, since the maximum fibre length has to be reduced to cater for the increased bandwidth, the minimum detectable current increases with $(bandwidth)^{3/2}$. Thus, for a 10MHz bandwidth a maximum sensitivity of 0.3A rms is projected for a 10mm coil. Above 100 MHz the minimum detectable current increases with $(bandwidth)^{5/2}$, since the detection noise is now dominated by the FET channel noise in the preamplifier.

The environmental robustness of a 100 turn prototype sensor has been assessed and found to have a measurement repeatability of ±0.5% and a temperature drift of 0.05%/°C; a consequence of using birefringent fibre. These current sensor are insensitive to external loading and can be wound as small as 13mm in diameter with only a 1 per cent reduction in sensitivity. The fibre is also found to be ~40dB less vibration sensitive than conventional fibre. The ability to wind small fibre coils while retaining high current sensitivity allows short fibre lengths to be used and high bandwidth are therefore possible. Typical performance gives a maximum current sensitivity of $1mA$ rms $Hz^{-\frac{1}{2}}$ and a large signal response of 450A rms, a dynamic range of 113dB. With minor improvements, a maximum sensitivity of $100\mu A$ rms $Hz^{-\frac{1}{2}}$ and dynamic range of 140dB might be reached and compact all-dielectric current sensors with milliampere sensitivity in the megahertz range can be projected.

2.2 Optical fibre gratings

Three main narrowband (optical fibre) filter types have emerged, with quite different characteristics. These are the fibre relief grating[9], photorefractive Bragg grating[10] and the miniature fibre Fabry-Perot[11] (FFP). Spectral filters consisting of concatenated twin-core fibre have also been demonstrated[12]. Extinction ratio and filter bandwidth are strongly dependent on waveguide geometry, the separation and refractive indices of the two cores. The relief grating and FFP are also modified waveguide structures whilst the photorefractive Bragg fibres rely on glass composition to allow a large, and stable refractive index change to be generated in the fibre core. Also, for pure simplicity and low cost, the rare-earth-doped fibre absorption filter[13] (described below) is hard to match.

2.2.1 Fibre Relief Grating Filters

So called because they apply a physical corrugation on the side of the fibre, fibre relief gratings are made by polishing the fibre cladding to obtain access to the core evanescent field, or employing a D-section fibre. Photoresist is then applied and exposed holographically using a short-wavelength laser. A grating is then etched into the fibre using either a dry or wet process. A high-index layer is then normally applied to "lift" the field and optimise its interaction with the grating corrugations.

Fibre relief gratings act either as a narrowband Bragg reflector, or as a band edge filter[14] in transmission. The latter characteristic is caused by radiation through the grating of wavelengths shorter than the Bragg-resonance, while longer wavelengths are transmitted. Reflectivities as high as 95 per cent with a bandwidth between 25 and 1,800GHz have been reported[15]. Excess losses are (<0.5dB) and limited tunability (3nm) can be obtained by temperature tuning or changing the index of the grating overlay. This process is suitable for all fibre compositions but has been superseded by the photorefractive Bragg grating which is more easily written.

2.2.2 Photorefractive Bragg Gratings

An exciting and relatively recent development is the technique of directly writing photorefractive Bragg gratings within the core of a single-mode fibre[10]. An interference pattern of ultraviolet light at around 240nm is focused onto the core of a germanium-doped silica fibre. After exposure for a few minutes, or even a single pulse, a distributed Bragg reflector is created at a wavelength corresponding to the periodicity of the interference pattern. Using this transverse holographic technique, photorefractive gratings can be written at any wavelength and reflectivities in excess of 99% have been demonstrated[16]. The origin of the effect is the subject of much controversy, with several theories available[17]. There appears to be agreement that it is associated with GeE' centres within the core which absorb strongly at 240nm but writing of the gratings is accompanied by an increase in absorption of only 0.2 per cent at $1.55\mu m$. Predictions of refractive index change at wavelengths far from the GeE' absorption band based on colour centre populations are two orders of magnitude smaller than the observed index changes. The colour-centre model should however be useful for calculating the induced space charge fields suspected to give rise to second-harmonic generation in optical fibres. Comparisons between as-pulled and annealed fibre suggest that the observed index changes may be predominantly due to compaction of the glass, the UV or green/blue light triggering stress relief in the glass by bond breakage. This mechanism may also explain the birefringence that is induced parallel to the exposing optical electric field. A simple model for changes in guided phase index with core radius yields ΔN_{eff} $\sim 10^{-4}$ for several tens of microstrain at 1550nm.

Unlike fibre relief gratings, photorefractive gratings act as a bandstop filter, reflecting the stopped light. Reflection linewidths of 20-100GHz are obtainable and the filters exhibit limited tunability (2nm) using either temperature or strain[18]. The main appeal of photorefractive grating filters is the ease with which they can be made, their very low loss and the non-invasive nature of the fabrication process.

2.2.3 Miniature Fibre Fabry-Perot Filters

These devices have been available for a number of years, but have recently come to prominence with the development of widely-tunable commercial devices. Several configurations are possible, the most popular being to deposit highly-reflective multi-layer dielectric mirrors on the ends of a short (<2mn) stub of fibre which is then glue-spliced between fibre pigtails. Tuning is achieved by piezo-electrically stretching the short fibre length, which incorporates a gap for this purpose. The inclusion of a fibre waveguide within the Fabry-Perot resonator is crucial, since it prevents beam walk-off and allows a high-finesse (>100) to be achieved in a compact, robust device.

FP Filters differ from grating filters in that they have a bandpass characteristic, reflecting the stopped light. In common with all Fabry-Perots, they exhibit multiple passbands, with typical bandwidth of 1-100GHz and a free spectral range (FSR) of 100-10,000GHz has been demonstrated. Excess losses are <3dB and tunability over one or more FSR is possible[19].

Photorefractive Bragg gratings are preferred for ease of writing and high performance, but are most successfully written in hosts containing GeO_2. Photo-induced refractive index changes in other silica-based optical fibre compositions, e.g Al_2O_3-SiO_2 are negligible. However it is found[20] that the addition of ~10,000ppm Ce^{3+} to the glass causes large index changes (~10^{-4}) under uv excitation. Similar effects have been observed[21] in heavily Eu_2O_3 doped phosphate glasses. The mechanism has not been fully explained and it is clear that the potential for grating formation has not yet been fully realised in fibres doped with metal ions.

3 MODIFIED MATERIALS

Many multicomponent glasses exhibit large nonlinear effects and can be drawn into single mode optical fibre via the rod-in-tube technique[22]. Features of particular interest include $\chi^{(2)}$, responsible for second harmonic generation, parametric interactions etc.[22] and $\chi^{(3)}$ responsible for the optical Kerr effect. Further, the host may be optimal for active, laser ions, allowing large emission cross sections and low concentration quenching, leading to, for example, efficient Q-switching (see below). However, it is the advent of rare-earth ion doping of conventional optical fibres[23] which has led to many developments in sensors and devices. This class of material offers many of the advantages of multicomponent glasses whilst retaining the practicality of, and compatibility with, silica. Nevertheless, many devices based on low phonon energy glasses such as ZBLAN have been demonstrated for efficient operation of systems which are dominated by non-radiative decay in silica. The 1.3μm amplifier[24] employing Pr^{3+} doped ZBLAN is currently the most attractive of this class although upconversion laser[25,26] and longer wavelengths (>2μm) systems (see, for example, reference 27) have attracted interest.

Passive effects demonstrated in rare-earth ion doped silica based fibres include photorefractive sensitivity as outlined above and optical filters using the intense narrow absorption bands arising from electronic transitions with the ion. Changes in absorption and emission spectra or decay time arise with temperature due to thermal population of the energy levels so heavily doped fibres may be employed as sensitive temperature monitors. Examples of the range of devices are described here.

The most successful fibre devices to date have been the active laser systems. The combination of waveguide geometry and rare-earth ion doped gain medium has proved particularly powerful with many totally new laser wavelengths demonstrated (for a recent review see reference 28). However, for sensor applications the potential for tunability and high power pulsed operation are most attractive, in for example remote gas sensors or distributed systems respectively. Examples of fibre designs for high performance laser operation in configurations suitable for sources in sensor systems are described here.

3.1 Fibre filter

This is the most straightforward of rare-earth ion doped fibre devices[13] and relies on the strong spectral variation in attenuation in the fibre due to the presence of the dopant ion. These absorption bands are intense but narrow and as a demonstration of the high extinction ratio achievable for small wavelength separations a length of Ho^{3+} doped fibre has been used to detect the anti-Stokes spontaneous Raman scattering signal from the pump wavelength, a differential intensity of $^-10^7$. The rejection ratio at a given wavelength is simply determined by the choice of dopant ion and dopant concentration or length of filter fibre spliced onto the output end of the fibre under test. The transmission of 633nm pump light and the signal generated on passing through 20m of Raman generating fibre and 7m of filter fibre are shown in Figure 3. The edge of the spontaneous anti-Stokes scattering is seen to dominate, indicating that transmission of the pump is around $3*10^8$ below the input level, without adversely affecting the Raman signal despite a wavelength separation of only 17nm.

3.2 Temperature sensor

3.2.1 Absorptive sensor

Changes in absorption spectra of metal ions occur with temperature as Stark components of the ground state energy level become populated according to the Boltzmann principle. Thus increased absorption on the long wavelength end of an absorption band can be correlated with temperature[29]. The change at 904nm for a GeO_2-SiO_2 core fibre doped with 5ppm Nd^{3+} has been determined to be linear over the range 100°C to -50°C with a sensitivity of 0.2%/°C. The upper limit is imposed by degradation of the acrylate fibre coating. The lower limit is determined by baseline loss, typically a few dB/km compared with 50dB/km absorption in this fibre at 904nm and 0°C. Nd^{3+} was chosen for its high temperature sensitivity at a convenient wavelength, but an extensive analysis of doped fibres[30] suggests that, at the operating wavelength of most commercial optical time domain reflectometry systems Cu^{2+} may be a more appropriate dopant.

The low dopant concentrations required for distributed sensing over long lengths and practical temperature ranges limit the spatial and temperature reduction achievable. Further, fluctuations in doping level will further limit measurement accuracy but early trials demonstrated performance of 2°C and 15m resolution over 140m.

An alternative approach is to employ heavily doped multimoded fibres as multiple point sensors[31]. An aluminosilicate core fibre of 0.2NA and 7.5wt% (27,000ppm) Nd^{3+} has been prepared by solution doping[32], giving an absorption of 3.8dB/cm at 904nm. Such fibres allow resolutions of a few centimetres and less than 1°C to be achieved in a multiple point configuration.

3.2.2 Fluorescence sensor

This sensor relies on a reduction in 1/e emission decay time with increasing temperature which is found to be linear[31] for an aluminosilicate sample doped with 3.5wt (18,400ppm) Nd^{3+} corresponding to a sensitivity of $0.25\mu s/°C$. The maximum dopant concentration is limited by concentration quenching[33] which causes a reduction in decay time. Although a reduction in lifetime with increasing concentration is recorded in aluminosilicate core fibres no fast decay component of the type found in germanosilicate core fibre is seen, suggesting aluminosilicates behave as multicomponent glasses with uniform dopant incorporation.

3.3 Fibre Lasers

The powerful combination of gain medium and waveguide geometry allows a wide range of laser sources to be constructed over an extensive wavelength range[28] with cw room temperature lasers having been demonstrated at wavelengths ranging from the blue to around $3\mu m$. Laser diode pumping is often possible due to the high power densities in the fibre core. Numerous fibres laser geometries have been constructed, in lengths ranging from a few millimetres to kilometres and fibre technology is sufficiently advanced that circuits can be manufactured for specific applications (a review is given in reference 34). These features, in conjunction with their simple construction and thermal stability, make fibre lasers practical and robust sources.

Linewidths of a few tens of kHz are obtained from ring laser cavities with integrated Fabry-Perot filters[35] whilst a figure eight combination leads to the generation of soliton pulses[36]. Conversely, by careful configurational and host design, stable broad bandwidth superfluorescent sources at $1.55\mu m$ and $1.06\mu m$ have been demonstrated[37,38] for use in gyroscopic applications.

Three examples are given here to indicate the range of designs achievable for use in sensor systems; a Q-switched source producing short high power pulses for OTDR, a tunable narrow linewidth laser for remote methane sensing and a high power cw single mode source giving extremely high output power densities.

3.3.1 Q-switched source

Q-switching of fibre lasers provides a means whereby the low CW power output from an inexpensive laser diode (used as a pump source) can be converted to high-power, short pulses suitable for use in ranging, OTDR and time-multiplexed fibre sensor systems. For a number of these systems the pulse duration sets the spatial resolution limit and hence short pulses of high power are required. Q-switched, laser diode pumped Nd:YLF lasers have been developed for such applications, providing Q-switched pulses as short as 2.7ns with 1.5kW peak power using a 200mW laser diode pump. Use of an amorphous host and fibre geometry gives[39] tunable operation over 40nm, around $1.053\mu s$, 2ns pulses of peak powers in excess of 1kW and repetition rates up to 1kHz, for an absorbed pump power of only 22mW at 810nm. These high powers and short pulse durations have been obtained by use of a special, high-gain neodymium-doped fibre and an electro-optic Q-switch as the modulator element. A heavily Nd^{3+} doped (1wt%) phosphate host is employed to give high emission cross sections and hence high gain efficiency whilst the relatively high dopant concentration of the fibre means that only a short fibre length is required to absorb the pump and this allows relatively short cavities to be constructed. With a high gain amplifier and a short cavity round-trip period, the net gain per unit time (dB/ns) experienced by signal photons in the cavity after Q-switch opening can be very high, even when large values of output coupling are used. This property enables short duration pulses to be obtained in the Q-switched configuration when a fast modulator is used as the Q-switch. Furthermore, using relatively high values of cavity output coupling allows efficient cavity energy extraction to be obtained by minimising the effect of unwanted cavity intrinsic losses.

The typical laser configuration is shown in Figure 4. A 25mm length of Nd^{3+}-doped phosphate fibre was employed and a dichroic dielectric coating with >99% reflection at $1.053\mu m$ and >95% transmission at 810nm was applied to one end, while a 1mm thick glass slide was bonded with index-matching adhesive to the other in order to displace the 4% Fresnel reflection from the waveguide end. This reflection thus occurs at a position where the beam is divergent and this reduces feedback into the amplifying fibre, so preventing premature oscillation of the high gain laser (typically 40dB round trip gain). An intro-cavity lens collimates the fibre output onto a 30% reflection mirror, providing the laser output and round-trip feedback. A 50mm long electro-optic, integrated (including polariser) Q-switch is used to modulate the cavity loss with extinction of >500:1 for 3.2kV applied voltage, with <1ns switch time capability. Although compact, the modulator still formed the major component in the overall cavity length (11cm). Typical results are shown in Figure 5. A 100mW single stripe laser diode operating at 810nm was employed to achieve these results, indicating the practical potential of this fibre laser configuration. Further, the reported tuning range of 40nm was achieved simply by replacing the output mirror with a reflection grating.

3.3.2 Tunable fibre laser

A tunable laser source is an attractive high-intensity source for high resolution spectroscopy. Such a source can be employed for the optical detection of gases by scanning it through one or more absorption lines of the gas[40]. Previously, semiconductor diode lasers have been demonstrated as possible sources for this application, but it has generally proved difficult to achieve high-yield manufacture of semiconductor lasers having a precise emission wavelength. However, rare-earth ions incorporated into amorphous media, eg, silica based optical fibre, exhibit broad linewidths absorption and emission due to inhomogeneous broadening. Further, laser emission can be tuned by incorporating a photorefractive grating in the laser cavity.

Rare-earth ion emission spectra are host sensitive thus the wavelength dependence of threshold shows a strong glass dependence. Data[41] are presented in Figure 6 for the three level system Tm^{3+} which exhibits both exceptionally broadband emission, ~300nm around 1700nm, and strong host sensitivity of emission. Further, broad absorption linewidths are advantageous for semiconductor laser diode pumping since only small (~10%) penalties in laser threshold occur over deviations of several nanometres from the optimum pump wavelength[40].

For use in gas sensing the laser is required to be tuned through an absorption line thus a laser cavity of conventional design is employed with an integral fibre Bragg grating used as the tuning element. For the detection of methane the emission is tuned through the P branch in the $2\nu_3$ absorption spectrum, around 1684nm. A dichroic mirror is butted to one end of the active fibre with a fibre photorefractive grating at the other. Stretching the fibre by heating with a Peltier unit tunes the grating reflection and hence the laser emission wavelength, as shown in Figure 7. Laser linewidths of less than 0.1nm are readily achieved (see insert). To demonstrate gas sensing the transmission through a cell containing 2.5% CH_4 in air is normalised to transmission through the cell when filled with N_2. Although the grating employed in this experiment was relatively poor (60%) a change of approximately 50% in normalised transmission is reported. An improved sensitivity is anticipated with better grating efficiencies and fully optimised laser cavity.

3.3.3 Cladding pumping

Careful combination of composition and geometry allows high power diode arrays sources to be exploited for high power cw single mode laser output[42]. The waveguide geometry, shown in Figure 8, typically comprises[43] a heavily rare-earth ion doped core (eg. 3wt% Nd^{3+}) which is located centrally in a rectangular, highly multimoded inner cladding waveguide into which pump light is injected. This is clad with a third material to give a high numerical aperture and circular fibre cross section. The inner cladding is designed to match the large diode diffraction angle and emitting area, whilst minimising the core/inner cladding area ratio, thus optimising the pump absorption in the core and minimising the laser threshold.

Multicomponent glass configurations have proved most successful to date because high doping levels, the required NA's and geometry can be achieved in an all-glass structure. Fibre losses tend to be high (0.1dB/m) so limiting length and requiring dopant concentrations of many wt%. Silica fibres offer the potential of longer devices due to lower outer waveguide losses (0.01dB/m) and hence reduced concentration quenching in the core but the silicone rubber or plastic outer cladding currently used is more difficult to handle.

Cladding pumped lasers and power amplifiers operating at 1.06μm and 1.535μm have been demonstrated. The Yb^{3+}/Er^{3+} codoped energy transfer configuration[44] has been shown to be a practical route to achieving low laser thresholds and high efficiency in the three level Er^{3+} system[45]. However, to date the most impressive results have been achieved in Nd^{3+} doped systems. Performance is comparable to that of the conventional geometry[43] but owing to the efficient use of high power diode arrays 4W cw single mode output at 1.06μm has recently been demonstrated[46], corresponding to power densities of order TW/m^2.

4. CONCLUSION

The optical fibre is shown to be substantially more versatile than just a passive light conduit. Devices based on passive sensitivity to specific measurements have been demonstrated and high performance tailor made active devices using all fibre circuitry are now under development as sources for senors systems.

Fibre design, both compositional and geometric, have proved critical in the achievement of these systems and with continued progress other practical devices are anticipated.

5. ACKNOWLEDGEMENTS

The Optoelectronics Research Centre is an SERC supported interdisciplinary research institute.

6. REFERENCES

1.   A.M.Smith, "Polarisation and magneto-optic properties of single mode optical fibre", Applied Optics, vol 17(1), pp.52-56, 1978

2.   H.O.Edwards, K.P.Jedrzejewski and R.I.Laming, "Optimal design of optical fibre for current measurement", Applied Optics, vol.28(11), pp.1977-1979, 1989

3.   C.C.Robinson, Applied Optics, vol.3, pp.1163, 1964

4.   R.I.Laming and D.N.Payne, "Electric current sensors employing spun highly-birefringent optical fibres", J. Lightwave Technology, vol.LT-7, pp.2084-2094, 1989

5.   D.N.Payne, A.J.Barlow and J.J. Ramskov-Hansen, "Development of low and high birefringence optical fibres", IEEE J. Quantum Electronics, Vol.QE-18(4), pp.477-488, 1982

6.   R.D.Birch, "Fabrication and characterisation of circularly birefringent helical fibres", Electronics Letters, vol.23(1), pp.50-52, 1987

7.   L.Jeunhomme and M.Monerie, "Polarisation maintaining single mode fibre cable design", Electronics Letters, vol.16, pp.921-922, 1980

8.   R.C.Jones, "A new calculus for the treatment of optical systems, Parts I-III", J. Optical Society of America, vol.31, pp.488-503, 1941

9.   W.V.Sorin and H.J.Shaw, "A single mode fibre evanescent grating reflector", J. Lightwave Technology, vol.LT-3(5), pp.1041-1043, 1985

10.   G.Meltz, W.W.Morey and W.H.Glenn, "Formation of Bragg gratings in optical fibres by a transverse holographic method", Optics Letters, vol.14(15), pp.823-825, 1989

11.   J.Stone and D.Marcuse, "Ultrahigh finesse fibre Fabry-Perot interferometers", J. Lightwave Technology, vol.LT-4, pp.382-385, 1986

12.   K.Okamoto and J.Noda, "Fibre optic spectral filters consisting of concatenated dual core fibres", Electronics Letters, vol.22(4), pp.211-212, 1986

13.   M.C.Farries, J.E.Townsend and S.B.Poole, "Very high rejection ratio optical fibre filters", Electronics Letters, vol.22(21), pp.1126-1128, 1986

14.   M.C.Farries, C.M.Ragdale and D.J.Reid, Proc 2nd Topical meeting on Optical Amplifiers and their applications, Paper ThD-1, Snowmass, Co., 1991

15.   I.Bennion, D.C.J.Reid, C.J.Rowe and W.J.Stewart, "High reflectivity monomode fibre grating filters", Electronics Letters, vol.22, pp.341-343, 1986

16.   L.Reekie, Private communication

17.   P.St.J.Russell, D.P.Hand, Y.T.Chow and L.J.Poyntz-Wright, "Optically-induced creation, transformation and organisation of defects and colour centres in optical fibres" Proc. SPIE vol.1516, International workshop on photo-induced self-organisation in optical fibres, Quebec, 1991

18.   W.W.Morey, "Distributed fibre grating sensors", Proc. Optical fibre sensors, pp.285, Sydney, 1990

19.   C.M.Miller, "A field-worthy, high performance, tunable fibre, fabry-perot filter", Proc. European Conference on optical communications, pp.605-608, Amsterdam 1990

20. L.Dong, P.J.Wells, D.P.Hand and D.N.Payne, "Uv-induced refractive index change in Ce^{3+} doped fibres", Proc. Conference on lasers and electro-optics, pp.68-71, Baltimore, 1991

21. E.G.Behrens, F.M.Durville and R.C.Powell, "Observation of erasable holographic gratings at room temperature in Eu^{3+} doped glasses, Optics Letters, vol.11(10), pp.653-655, 1986

22. E.R.Taylor, D.J.Taylor, L.Li, M.Tachibana, J.E.Townsend, J.Wang, P.J.Wells, L.Reekie, P.R.Morkel and D.N.Payne, "Application-specific optical fibres manufactured from multicomponent glasses, Proc. Materials Research Symposium on Optical fibre materials and processing, vol.172, pp.321-327, Boston, 1989

23. S.B.Poole, D.N.Payne and M.E.Fermann, "Fabrication of low loss optical fibres containing rare-earth ions", Electronics Letters, vol.21, pp.737-738, 1985

24. Y.Ohishi T.Kanamori, T.Kitagawa, S.Takahashi, E.Snitzer and G.H.Seigel, "Pr^{3+}-doped fluoride fibre amplifier operating at $1.3\mu m$", Optics Letters, vol.16(22), pp.1747-1749, 1991

25. R.G.Smart, J.N.Carter, A.C.Tropper, D.C.Hanna, S.T.Davey, S.F.Carter and D.Szebesta, "Cw room temperature operation of praseodymium doped fluorozirconate glass fibre lasers in the blue, green and red spectral regions", Optics Communications, vol.86(3-4), pp.337-340, 1991

26. J.Y.Allain, M.Monerie and H.Poignant, "Room temperature CW tunable green upconversion holmium fibre laser", Electronics Letters, Vol.26, pp.261-262, 1990

27. M.C.Brierley and P.W.France, "Continuous wave lasing at $2.7\mu m$ in an erbium doped fluorozirconate fibre", Electronics Letters, vol.23, pp.329-331, 1987

28. D.N.Payne, "Active fibres and optical amplifiers", AGARD, EPP/GCP Lecture no. 184 NATO Series of lectures, Rome, Amsterdam and Montery, May 1992

29. M.C.Farries, M.E.Fermann, R.I.Laming, S.B.Poole and D.N.Payne, "Distributed temperature sensor using Nd^{3+} doped optical fibre", Electronics Letters, vol.22(8), pp.418-419, 1986

30. Y.Suetsugu, S.Ishikawa, T.Kohgo, H.Yokota and S.Tanaka, "Temperature dependence of the absorption loss of the silica based optical fibres doped with transition and rare-earth metal ions", Proc. IOOC, Gothenburg, 1989

31. P.L.Scrivener, P.D.Maton, A.P.Appleyard and E.J.Tarbox, "Fabrication and properties of large core, high NA, high Nd^{3+} content multimode optical fibres for temperature sensor applications", Electronics Letters, vol.26(13), pp.872-873, 1990

32. J.E.Townsend, S.B.Poole and D.N.Payne, "Solution doping technique for fabrication of rare-earth doped optical fibres", Electonics Letters, vol.23(7), pp.329-331, 1987

33. K.Arai, H.Namikawa, K.Kumate, T.Honda, Y.Ishii and T.Handa, "Aluminium or phosphorous codoping effects on the fluorescence and structural properties of neodymium doped silica fibres", J. Applied Physics, vol.59(10), pp.3430-3436, 1986

34. P.Urquhart, "Review of rare-earth doped fibre laser and amplifiers", IEEE Proceedings Part J, vol.135, pp.385-406, 1988

35. P.R.Morkel, G.J.Cowle and D.N.Payne, "Travelling wave erbium fibre ring laser with 60kHz linewidth", Electronics Letters, vol.26(10), pp.632-634, 1990

36. I.N.Duling, "All-fibre ring soliton laser modelocked with nonlinear mirror", Optics Letters, vol.16(8), pp.539-541, 1991

37. P.R.Morkel, "Erbium doped fibre superfluorescent sources for the fibre gyroscope", Proc. Optical fibre sensors, vol.44, pp.143-148, Paris, 1989

38. P.R.Morkel, K.P.Jedrzejewski, E.R.Taylor and D.N.Payne, "High-gain superfluorescent neodymium doped single mode fibre source", IEEE Photonics Technology Letters, vol.4(7), pp.706-708, 1992

39. P.R.Morkel, K.P.Jedrzejewski, E.R.Taylor and D.N.Payne, "Short pulse high power Q-switched fibre laser", IEEE Photonics Technology Letters, vol.4(6), pp.5454-547, 1992

40. W.L.Barnes, J.P.Dakin, H.O.Edwards, L.Reekie, J.E.Townsend, S.Murray and D.Pinchbeck, "Tunable fibre laser source for methane detection at $1.68\mu m$", SPIE OE/Fibres, Boston, 1992

41. W.L.Barnes and J.E.Townsend, "Highly tunable and efficient diode pumped operation of Tm^{3+} -doped fibre lasers", Electronics Letters, vol.26(11), pp.746-747, 1990

42. E.Snitzer, H.Po, F.Hakimi, R.Tumminelli and B.C.McCollum, "Double-clad offset core Nd fibre laser", Proc Conference on Optical Fibre Communications, New Orleans, Paper PD5, 1989

43. J.D.Minelly, E.R.Taylor, K.P.Jedrzejewski, J.Wang, D.N.Payne, "Laser diode pumped neodymium doped fibre laser with output power in excess of 1Watt", Proc. Conference on lasers and electro-optics, Paper CWe6, Anaheim , 1992

44. J.E.Townsend, W.L.Barnes and S.G.Grubb, "Yb^{3+} sensitised Er^{3+} doped, silica based optical fibre with ultra-high transfer efficiency", Materials Research Symposium on Optical waveguide materials, vol.244, pp.143-146,1991

45. J.D.Minelly, W.L.Barnes, P.R.Morkel, J.E.Townsend, S.G.Grubb and D.N.Payne, "High power Er^{3+}/Yb^{3+} fibre laser pumped by a 962nm diode array", Proc. Conference on lasers and electro-optics, Paper CPD-17, pp.33-34, Anaheim, 1992

46. J.D.Minelly Private communication

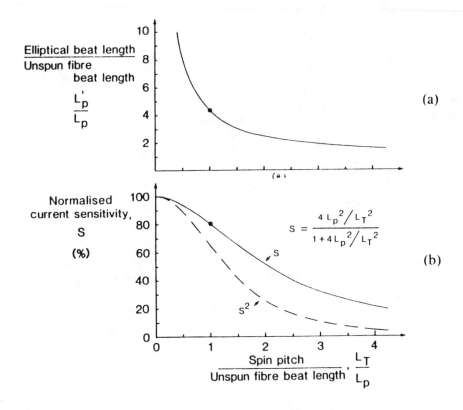

Fig. 1 a) Normalised elliptical beat length and b) relative current sensitivity as a function of the ratio of spin pitch to unspun fibre beat length.

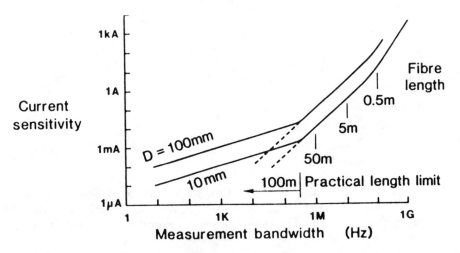

Fig. 2 Calculated current sensitivity of SEB fibre current sensors assuming the maximum fibre length permitted for a given bandwidth is used. The curve is shown for 10mm and 100mm fibre coils and is restricted to a practical fibre length of 100m.

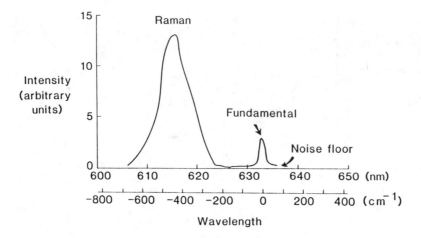

Fig. 3 Relative pump and signal levels after passing through 20m of Raman generating fibre and 7m of filter fibre showing pump rejection and Raman transmission.

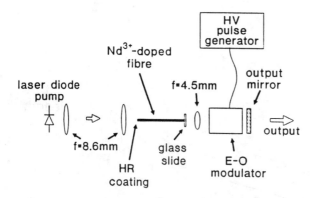

Fig. 4 Q-switch fibre laser configuration with integrated electro-optic Q-switching element. The glass slide reduces feedback into the amplifying fibre and prevents premature oscillation of the laser.

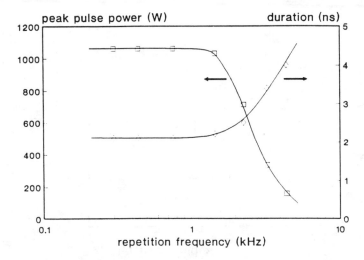

Fig. 5 Dependence of Q-switched pulse peak power and FWHM duration on repetition frequency for maximum pump diode output power of 89mW (22mW absorbed).

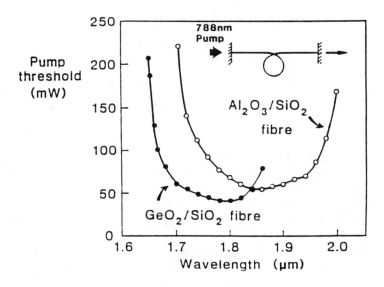

Fig. 6 Tuning curves for Tm^{3+} doped fibre lasers in GeO_2-SiO_2 or Al_2O_3-SiO_2 core fibres, with grating as output coupler.

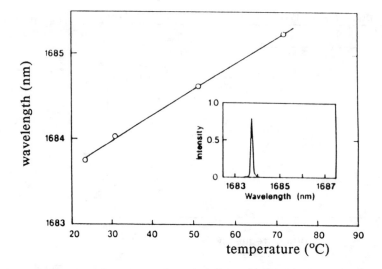

Fig. 7   Temperature tuning of Tm^{3+} doped GeO_2-SiO_2 core fibre achieved by Peltier heating a fibre Bragg grating integral in the laser cavity

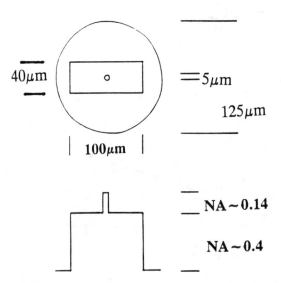

Fig. 8   Schematic showing the fibre geometry and refractive index profile for a typical cladding pumped fibre

SESSION 2

Discrete Fiber Optic Sensors

Chair
Gordon L. Mitchell
MetriCor, Inc.

Etalon-based Fiber Optic Sensors

C. E. Lee and H. F. Taylor
Department of Electrical Engineering
Texas A&M University
College Station, Texas 77843 U. S. A.

ABSTRACT

Progress in the development of fiber sensors with the Fabry-Perot interferometer (etalon) configuration is reviewed. Fabrication methods and performance results for intrinsic and extrinsic sensors are presented. Application of these devices in measuring temperature, strain, ultrasonic pressure, and gas pressure in internal combustion engines is described. Techniques by which the fiber Fabry-Perot sensors have been successfully embedded in composites and metals are indicated.

1. INTRODUCTION

The Fabry-Perot interferometer (FPI), sometimes called the Fabry-Perot etalon, consists of two mirrors of reflectance R_1 and R_2 separated by a distance L, as in Fig. 1. Since its invention in the late 19th century [1], the bulk-optics version of the FPI has been widely used for high-resolution spectroscopy. In the early 1980s, the first results on fiber optic versions of the FPI were reported. In the late 1980s, fiber Fabry-Perot interferometers (FFPIs) began to be applied to the sensing of temperature, strain, and ultrasonic pressure in composite materials.

The FFPI would appear to be an ideal transducer for many smart structure sensing applications. As with all fiber interferometers, the FFPI is extremely sensitive to environmental perturbations. The sensing region, which consists of the portion of the fiber core (or in some cases an air gap) between the two mirrors, can be very compact - equivalent to a "point" transducer in many applications. Since it is an electrical insulator, it is not affected by electromagnetic interference. Unlike other fiber interferometers (Mach-Zehnder, Michelson, Sagnac) used for sensing, the FFPI contains no fiber couplers - components which can greatly complicate the deployment of the sensor and the interperetation of data it produces. Finally, FFPI sensors are amenable to both time-division multiplexing and coherence multiplexing.

2. CONSTRUCTION AND OPERATION OF THE FFPI SENSOR

The FFPI configuration of Fig. 1 makes use of internal mirrors - the mirrors are an integral part of a continuous length of fiber. Internal mirrors formed from dielectric coatings have shown the best mechanical properties, lowest excess optical loss, and widest range of reflectance values. The most commonly used mirror material is TiO_2, which has a refractive index of 2.4 (vs. 1.46 for fused silica) [2,3]. The reflection results from the refractive index discontinuities at the two film-fiber interfaces. The TiO_2 films have been produced by sputtering in an rf planar magnetron system, or by electron beam evaporation. Typical film thicknesses are in the neighborhood of 1000 Å. The fusion splicer is operated at lower arc current and duration than for a normal splice, and several splicing pulses are used to form a mirror. The mirror reflectance generally decreases monotonically as a function of the number of splicing pulses, making it possible to select a desired reflectance over the range from much less than 1% to about 10% during the splicing process. To achieve internal mirror reflectances

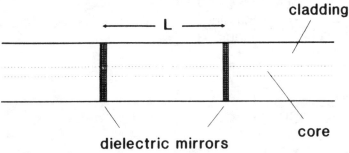

Fig. 1. Fiber Fabry-Perot interferometer with two internal mirrors.

greater than 10%, multilayer TiO_2/SiO_2 films produced by magnetron sputtering have been used. The best result to date with multilayer mirrors is a reflectance of 86% in an FFPI with a finesse of 21 at a wavelength of 1.3 μm [4].

Most FFPI sensors investigated to date have made use of low-reflectance mirrors. Assuming that the mirrors are lossless and have equal reflectances, with $R = R_1 = R_2$, it is easily shown that if $R \ll 1$, then

$$R_{FP} \cong 2R(1 + \cos \phi) \tag{1}$$

where R_{FP} represents the ratio of the power reflected by the FFPI to the incident power and ϕ, the round-trip phase shift in the interferometer, is given by

$$\phi = \frac{4\pi nL}{\lambda} \tag{2}$$

with n the refractive index of the region between the mirrors and λ the free-space optical wavelength.

An experimental arrangement to monitor the reflectance of an FFPI sensor is shown in Fig. 2.

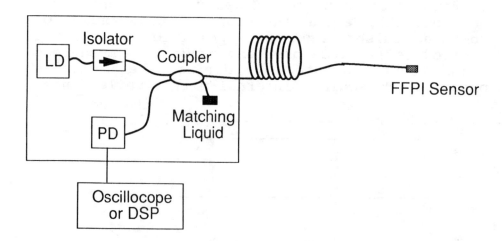

Fig. 2. Arrangement for monitoring the reflected power from a FFPI.

Light pulses from a 1.3 μm fiber-pigtailed laser
diode pass through an optical isolator to a fiber
coupler and are reflected by the FFPI. After
passing through the coupler again, the reflected
pulses are converted by a photodiode to an
electrical signal which is displayed upon an
oscilloscope. Some typical waveforms obtained in
this manner are shown in Fig. 3. If the laser
pulse is short, the driving current amplitude is
small, and the FFPI cavity is short, the
reflected amplitude is almost constant for the
duration of the pulse, as in Fig. 3 (a).
Increasing the pulse width, driving current
amplitude, and interferometer length causes the
reflected signal from the FFPI to vary with time
due to chirping of the laser. As the laser heats
during the pulse, its frequency and consequently
the reflected power from the interferometer
changes with time, as in Fig. 3 (b). Thus, the

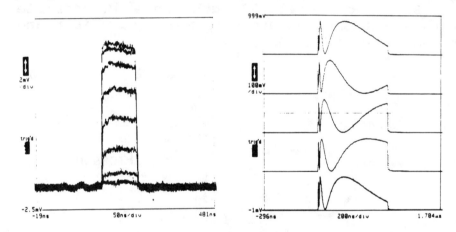

Fig. 3. Oscilloscope traces showing reflected
power from FFPI at different temperatures for a
pulsed laser input. In (a), the temperature was
changed from 77°C to 56°C in 3°C increments. The
interferometer was short (1.5 mm), the pulse
width was short (100 nsec), and the modulating
pulse amplitude was small so that the effect of
laser chirping was small. In (b), the
temperature was changed from 29°C to 23°C in
1.5°C increments. A longer (1 cm)
interferometer, longer pulse duration (0.8 μsec),
and larger modulating pulse amplitude causes the
interferometer output to sweep through fringes in
response to chirping of the laser.

temporal variation in the reflected pulse amplitude represents a sweeping through fringes of the interferometer.

The FFPI sensors described above make use of a fiber propagation path between the mirrors, and thus sense parameters which change the refractive index or length of the fiber. These are termed intrinsic Fabry-Perot sensors. Other sensors where the propagation path between the mirrors is "external" to the fiber are termed extrinsic sensors. One example of an extrinsic FFPI sensor uses a cavity formed by the fiber end and a membrane located beyond the fiber end [5,6], illustrated in Fig. 4. The reflectance from the interferometer changes in response to a motion of the membrane which affects the length of the external cavity. Typical separation between the membrane and the fiber end is of the order of 1 μm. Such short cavity lengths make it possible to operate these sensors with multimode fiber and low-coherence LED light sources.

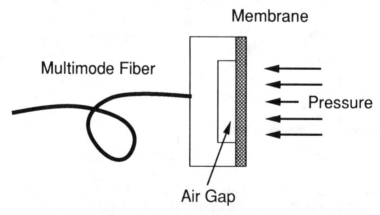

Fig. 4. Extrinsic FFPI sensor with a cavity formed by the air gap between the fiber end and a membrane.

3. APPLICATIONS OF THE FFPI SENSOR

The most common applications of the FFPIs have been in the measurement of temperature, strain, and ultrasonic pressure, as described below.

An FFPI with the configuration of Fig. 1 with internal mirror reflectances of about 2% and excess losses of 0.1 to 0.2 dB per mirror was used to sense temperature over the range -200^{0}C to 1050^{0}C [3]. No hysteresis or change in the mirror reflectances were observed over the temperature range of the experiment, although it was noted that the fiber became brittle at the higher temperatures. The sensitivity is lowest at cryogenic temperatures and the response is fairly linear at room temperature and above.

The conventional way to monitor strain in a structural part is to bond electrical strain gauges to its surface. The high sensitivity of the FFPI to longitudinal strain can be utilized in a similar manner. In the first experiment of this type, a FFPI with a cavity formed with a silver internal mirror and a silvered fiber end was bonded to the surface of a cantilevered aluminum beam [7]. A conventional resistive foil strain gauge was also bonded to the beam as a reference. Strain was introduced by loading the beam in flexure. The optical phase change in the FFPI, determined by monitoring the reflected power and counting fringes, was found to be a linear function of the strain reading from the resistive device over the range 0 - 1000 $\mu\varepsilon$.

The high strain sensitivity of FFPI sensors has also been utilized in the measurement of gas pressure in internal combustion engines [8]. Many engines are constructed such that an element such as a fuel injector valve which is exposed to the combustion chamber pressure is bolted to the cylinder head. The variation in longitudinal strain on the bolts during a engine cycle is approximately proportional to the combustion chamber pressure. An FFPI epoxied into a hole drilled in one of these bolts can thus be used for engine pressure measurements. Data obtained with a diesel test engine comparing the response of a FFPI with that of a conventional pressure transducer is shown in Fig. 5. Unlike the piezoelectric sensors presently used to monitor engine pressure, the FFPI does not need cooling and should operate for long periods of time without recalibration.

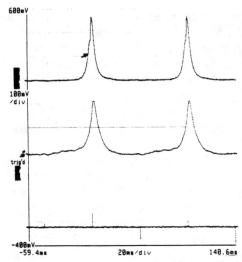

Fig. 5. Combustion chamber pressure in a diesel test engine as measured with a conventional piezoelectric pressure sensor (upper trace) and a FFPI (middle trace). The lower trace indicates the "top dead center" piston position.

Extrinsic FFPI sensors with the configuration of Fig. 4 have been use to measure temperature and pressure. The membrane is designed such that changes in the measurand cause a deflection of the membrane, thus affecting the length of the cavity.

Surface-mounted extrinsic FFPIs with the configuration of Fig. 6 have also been used for monitoring temperature, strain, and acoustic pressure [9]. In this case, an air gap

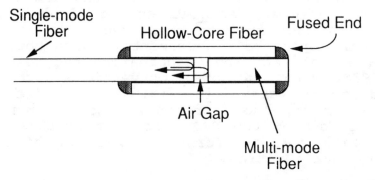

Fig. 6. Extrinsic FFPI with cavity formed by the air gap between the ends of a single mode and a multimode fiber.

separating the cleaved ends of a single mode fiber and a multimode fiber forms the FFPI cavity, typically a few μm to a few hundred μm in length. The fiber alignment is maintained by a silica tube in which the fibers are inserted. Each of the fibers is free to move longitudinally in the region near the cavity, but is constrained at some point along its length by bonding either to the silica tube or to the sample being monitored. The distance between bonding points is the length of the region over which a perturbation affects the sensor output, known as the gage length. The FFPI is monitored in reflection, with the input and reflected light carried by the single mode fiber.

In one experiment, two of the the extrinsic FFPIs were bonded on the surface of a ceramic material in close proximity to one another [9]. The cavity lengths in the two sensors were slightly different, so that the round-trip phase shifts were different by about $\pi/2$ radians. By simultaneously monitoring the two sensor outputs, it is possible to determine the direction of change of the phase shift and avoid sensitivity nulls. When the ceramic was temperature-cycled from 25^0C to 600^0C, the sensor data provided information on temperature change and on the expansion of a crack in the material.

In another experiment, an extrinsic FFPI bonded to the surface of an aluminum block was used to detect surface acoustic waves launched by a piezoelectric transducer at a frequency of 1 MHz [10]. A dual-wavelength scheme with lasers at .78 μm and .83 μm was used to obtain in-phase and quadrature signals from the sensor. The FFPI also detected the acoustic noise burst produced by breaking a pencil lead on the surface of the aluminum sample.

4. EMBEDDED FFPI SENSORS

Recently, there has been considerable interest in the embedding of fiber sensors in structural materials. A graphite epoxy composite was the material used in the first experiments on embedding of FFPIs [11]. The temperature sensitivity of the embedded FFPI was tested by monitoring fringes at a wavelength of 1.3 μm as the panel was heated from room temperature to 200°C. The value of the quantity Φ_T, defined as

$$\Phi_T = d\phi/\phi dT \qquad (3)$$

was measured to be 8.0 x 10^{-6}/ °C, slightly less than the value 8.3 x 10^{-6}/ °C for the same FFPI in air prior to embedding. From this data, a thermal expansion coefficient of 2.1 x 10^{-7}/ °C was estimated for the composite.

The ability to embed FFPIs in structural materials suggests application in as a pressure transducer in ultrasonic nondestructive testing (NDT). In conventional NDT studies, piezoelectric transducers (PZTs) are positioned on the surface of the sample to launch and detect ultrasonic waves. The ability to locate the receiving transducer deep within the material opens the possibility of obtaining new information on the properties of a bulk sample. For example, it may be possible to better detect and locate delamination sites in composites or cracks in metals. Experiments with a PZT transducer positioned on the surface of the composite sample described above yielded an acoustic response from the FFPI over the frequency range 100 kHz to 5 MHz [12]. The compressional acoustic wave generated by the transducer interacts with light in the fiber to produce a phase shift via the strain-optic effect. Only the acoustic wave in the region between the mirrors contributes to the sensor output.

Recently, the sensing of temperature and of ultrasonic pressure with FFPIs embedded in aluminum was demonstrated [13]. The aluminum parts were cast in graphite molds in air, as illustrated in Fig. 7. Breakage of the fibers at the air-metal interface during the casting process was avoided by passing the fiber through stainless-steel stress-relief tubes, which extended a short distance into the finished part. Thermal expansion of the aluminum caused the optical phase in the embedded FFPI to be 2.9 times more sensitive to temperature than for the same interferometer in air. The same FFPI was also been used to detect ultrasonic waves launched by a surface-mounted PZT transducer over the frequency range from 0.1 - 8 MHz.

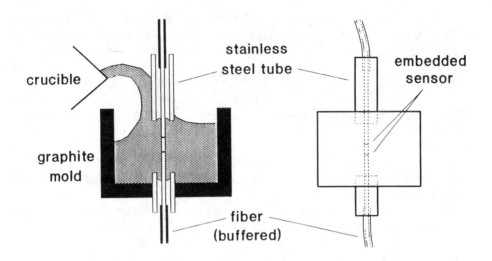

Fig. 7. Process for embedding FFPI in a cast aluminum part.

5. CONCLUSIONS

The fiber Fabry-Perot interferometer (FFPI) is a strong candidate for many sensing applications because it is extremely sensitive, provides point sensing capability, is immune to electromagnetic interference, is simple to produce, has excellent mechanical properties, and is amenable to time-division and coherence multiplexing. FFPIs have been used to sense strain, temperature, ultrasonic pressure, and gas pressure in an internal combustion engine. They have been embedded in composites and in aluminum, where they have sensed temperature, strain, and ultrasonic pressure. Further advances in manufacturing methods and signal processing to reduce the cost of individual sensing elements will be needed before widespread commercial application of FFPIs can be realized.

6. REFERENCES

1. C. Fabry and A. Perot, *Ann. Chim. Phys.*, **16**, 115 (1899).

2.   C. E. Lee and H. F. Taylor, Interferometric Optical Fibre Sensors using Internal Mirrors, *Electron. Lett.* **24**, 193 (1988).

3.   C. E. Lee, R. A. Atkins, and H. F. Taylor, Performance of a Fiber-Optic Temperature Sensor from - 200 to 1050°C, *Opt. Lett.* **13**, 1038 (1988).

4.   C. E. Lee, W. N. Gibler, R. A. Atkins, and H. F. Taylor, In-Line Fiber Fabry-Perot Interferometer with High-Reflectance Internal Mirrors, *J. Lightwave Technol.* (1992).

5.   R. A. Wolthuis, G. L. Mitchell, E. Saaski, J. C. Hartl, and M. A. Afromowitz, Devellopment of Medical Pressure and Temperature Sensors Employing Optical Spectrum Modulation, *IEEE Trans. Biomed. Engineering* **38**, 974 (1991).

6.   G. L. Mitchell, Intensity-Based and Fabry-Perot Interferometer Sensors, (*Fiber Optic Sensors: An Introduction for Engineers and Scientists*, E. Udd, Ed., Wiley, New York, 1991) p. 139

7.   T. Valis, D. Hogg, and R. M. Measures, Fiber Optic Fabry-Perot Strain Gauge, *IEEE Photonics Technol. Lett.* **2**, 227 (1990).

8.   R. A. Atkins, W. N. Gibler, C. E. Lee,M. Oakland, M. Spears, V. Swenson, H. F. Taylor, J. McCoy, and G. Beshouri, Fiber Optic Pressure Sensor for Internal Combustion Engines, to be published.

9.   K. A. Murphy, C. E. Kobb, A. J. Plante, S. Desu, and R. O. Claus, High Temperature Sensing Applications of Silica and Sapphire Optical Fibers, *Proc. SPIE* **1370**, 169 (1990).

10.   T. A. Tran, W. V. Miller III, K. A. Murphy, A. M. Vengsarker, and R. O. Claus, *Proc. SPIE* **1584**, 178 (1991).

11.   C. E. Lee, H. F. Taylor, A. M. Markus, and E. Udd, Optical-Fiber Fabry-Perot Embedded Sensor, *Opt. Lett.* **14**, 1225 (1989).

12.   J. J. Alcoz, C. E. Lee, and H. F. Taylor, Embedded   Fiber-Optic   Fabry-Perot   Ultrasound Sensor, *IEEE Trans. Ultrasonics, Ferroelectrics, and Freq. Control* **37**, 302 (1990).

13.   C. E. Lee, W. N. Gibler, R. A. Atkins, J. J. Alcoz, and H. F. Taylor, Metal-Embedded Fiber Optic Fabry-Perot Sensors, *Opt. Lett.* **16**, 1990 (1991).

Micromachined fibre optic sensors

Brian Culshaw

University of Strathclyde, Department of Electronic and Electrical Engineering,
Royal College Building, 204 George Street, Glasgow G1 1XW, UK

ABSTRACT

Micromachined sensors, especially those based upon silicon technology are compatible with simple batch processing and so, in principle, may be fabricated repeatably and economically. For relatively precise ($\simeq 0.1\%$ and better) measurements, microresonators in which the measurand is caused to modulate the resonant frequency of a mechanical structure can be particularly attractive. Micromechanics enables these resonators to be readily fabricated to the appropriate sub-millimetre dimensions which are compatible with all optical excitation and detection of the resonance phenomena. The ability to transmit the necessary optical signals along fibres results in a precise miniature and potentially rugged transducer system. These fibre optic resonator devices which are entirely electronically passive are particularly suited to applications involving environments subject to high electro-magnetic fields and/or to extremes of temperature.

This paper introduces the concepts underpinning the realisation of micromachined resonator fibre optic sensors, discusses the achievements in prototype sensor systems realised to date and projects the future requirements which must be fulfilled before the technology can realise its full commercial potential. Measurements to 0.1% accuracy over very wide temperature ranges have been demonstrated to be feasible and applications in the difficult environments typified by aerospace, industrial, and automotive measurements and in chemical/biochemical systems are emerging.

1. INTRODUCTION

Mechanical resonators have been used for many years as the basis of a wide range of sensors in which the measurand is caused to modulate the resonant frequency of the resonator in a carefully controlled and predetermined relationship[1]. The mechanical resonator is typically a wire or a cylinder and its motion is initiated and detected electrically. The principal motivations underlying the interest in resonator technology are that the measurement is represented by a specific frequency within a known range and this frequency is relatively straightforward to detect and measure with extremely high accuracy. Provided that the measurand is repeatably modulated on to the resonator and, in particular, that the effects of temperature variations can be neglected, the precision in the measurement of frequency is matched by the implicit precision in the specification of the measurand. Additionally the transmission of a frequency signal over most types of transmission line whether optical or electrical will not incur any distortion penalties within the transmission path no matter how this path is varied.

Resonators then are extremely attractive sensors. When these features are combined with a fibre optic feed and return link, then additional advantages such as the ability to work in high electrical and magnetic field environments, to withstand immersion in chemically active media and to withstand extremes of temperature also come into play. The last of these is a direct consequence of the removal of all electronics from the sensor head - in virtually all other technologies, the electronics within the sensor head limits the allowed temperature excursions especially at the high end of the range.

For this approach to be effective the signals transmitted to and returning from the sensor using the fibre optic link should remain in the optical domain and in particular optical to electronic conversion within the sensor head - whilst extremely convenient - implicitly removes some of the compelling advantages of this technology. Consequently, direct conversion from optical to mechanical energy is necessary within the sensor itself and some form of position tolerant optical detection of the consequent mechanical motion is required. It is desirable that the optical power levels transmitted to the sensor head should be relatively small (preferably fractions of a mW) so that small structures are required to optimise the effects of the optical to mechanical conversion process.

Silicon micromachining affords an ideal technology whereby such small precisely defined structures may be realised[2]. This paper examines the amalgamation of micromechanical resonators, fibre optics and mechanical and transducer engineering to realise a range of potentially valuable sensing elements with applications in aerospace, automobile and industrial control and biomedical and medical engineering. Indeed a number of trial structures have already been assessed as field prototypes and the results have been extremely encouraging.

The remainder of this paper is divided into three major sections. The first considers resonators and their use as transducers and goes on to examine the principles whereby silicon microresonators may be designed and fabricated. The second section examines the optics involved in exciting and observing the mechanical motion of microsresonators and finally some examples of prototype sensor configurations are discussed.

The proof of principle for optically excited and interrogated microresonator sensors is well established. There do, however, remain a number of important engineering issues such as the effects of temperature and the design of mechanically compatible launch and receive optics. These issues must be resolved before optically excited and interrogated microresonators emerge as production engineered measuring instruments. The approach to resolving these issues has been defined so that the future outlook is extremely optimistic. These small, precise yet rugged transducers will permit what are, as yet, intractable measurements and as such will make a significant contribution to the evolution of future transducer technology.

2. RESONATORS AND MICRORESONATORS

The basic elements of mechanical resonator systems are extremely simple and may be exemplified by discussing a simple one dimensional equation[3] :

$$F(t) = m\frac{d^2x}{dt^2} + \mu\frac{dx}{dt} + kx \qquad\qquad 1$$

where F(t) is an applied force as a function of time t, m is an (effective) mass, μ a "mechanical resistance" and k an (effective) stiffness. The force F is directed along the the direction of x and x is the displacement in that direction referred back to an equilibrium position when $F = 0$.

We may derive some very basic relationships from equation 1 and in particular :

$$\omega_0^2 = \frac{k}{m} \qquad\qquad 2a$$

$$Q = \frac{\omega_0 \, m}{\mu} = \frac{k}{\omega_0 \mu} \qquad \qquad \text{2b}$$

where ω_0 is the mechanical resonant frequency and Q the Q factor.

For a generalised three dimensional system both m and k are typically directionally dependent so that the resonant frequency spectrum is different for different excitation conditions. This in turn implies that the excitation conditions must be arranged such that any driving force is applied *only* along one of the principal axes of the system, otherwise resonant frequency ambiguities can, and indeed often do, occur. Furthermore, m and k for a given excitation direction are also functions of a mode number and again the excitation direction is best chosen so that resonant frequencies corresponding to particular successive mode numbers are well separated. These basic and very simple observations have a profound effect upon the design of mechanically resonant transducers.

In addition to the resonant frequency the Q factor is important and in principle the higher this is the simpler it is to determine the value of ω_0 for a particular resonator since Q is the reciprocal of the normalised bandwith. There are some disadvantages to extremely high Q systems (and mechanical resonators with Q's over half a million have been reported) and in particular the driving frequency needs to be within, broadly, ω_0/Q of ω_0 for excitation to commence. This need for exact matching imposes, as the Q increases, the necessity for self starting oscillation systems which use the mechanical element as the frequency determing element. The higher the Q the higher also is the build up time from zero displacement to a dynamic equilibrium. This is of the order of Q cycles. So, for example, an oscillator with a resonant frequency of 100kHz and a Q of a half a million will take 5 seconds to stabilise from switch on. If the value of ω_0 is changed after equilibrium has been established, for example by altering the effective stiffness, the mechanical oscillator will follow the new resonant frequency effectively immediately and will *not* take the Q cycles to establish new resonant regime. Classical, rock and blues guitarists all exploit this to great effect by running their fingers down the frets after the note has been struck. Incidentally, in this case, it is the effective mass of the resonator which is being changed rather than its effective stiffness which is dominated by the tension applied in the string.

This brings us on to the topic of resonators as transducers and Equation 1 and 2 say it all. If the measurand is to affect ω_0 then either k or m must change. For most mechanically resonant transducer systems, the measurand affects k usually by virtue of applying tension (or compression) to the resonant element. The value of m can only be altered by absorption (or sublimation) of material to or from the resonator, or by changing the physical dimensions of the resonator. Usually these interaction mechanisms are extremely inconvenient.

For a particular resonator oscillating in a particular mode the resonant frequency is conveniently viewed as the sum of two terms :

$$\omega_o^2 = \omega_M^2 + \omega_T^2 \qquad \qquad 3$$

where ω_M is the resonant frequency of the unstressed material forming the reonator and ω_T is a term describing the influence of applied stresses on the resonator. Typically ω^2_T is linearly dependent upon the magnitude of applied stress.

Most of the experimental work described in this paper centres upon resonant characterstics of a simple flat rectangular beam (Fig.1) of thickness t and width w and length L clamped to both ends and operating in its fundamental mode. For such structures it is relatively straightforward to show that :

$$f_1 = \frac{\omega_{M1}}{2\pi} = \frac{1.028t}{L^2} \left(\frac{E}{\rho}\right)^{\frac{1}{2}}$$ 4

where f_1 is the fundamental resonant frequency, E is the Young's Modulus of the material which forms the bar and ρ its density. The oscillations in this case occur in the Y direction as indicated in Figure 1. For most of the structures considered, the ratio of w to t is quite large so there will be a significant difference in the resonant frequencies for excitation in the two directions. Furthermore, the next resonant frequency for the second order mode f_2 has a value of 1.66 f_1 and also has even as opposed to odd modal symmetry about the centre of the beam and for both these reasons is unlikely to be excited.

The effect of an applied longitudinal tension T on the resonant frequency can be derived as :

$$f_0^2 = f_1^2 + 0.318 \frac{T}{L^2 wt\rho}$$ 5

It is interesting to note that equation 5 tends to the equation describing an infinitely flexible string (ie one for which E = 0) as the value of E and consequently the value of f_1 tends to zero. Thus the original assertion in equation 3 is borne out by more detailed analysis.

The specific properties of silicon microresonators may be readily related back to these basic equations. In the sketch of a typical microresonator assembly shown in Figure 2, the resonator itself is defined by using a boron doped layer as an etch stop and undercutting this layer using isotropic etch techniques. Typical dimensions are shown in the diagram and the basic anisotropic etched based fabrication processes have been described extensively in the literature[2].

The influence of the presence of boron atoms on the resonant frequency behaviour of this type of resonator is extremely important. A successful etch stop requires the presence of approximately 0.1% boron. The boron atom is about the half the size of the silicon atom, so displacing one atom in a thousand in the silicon lattice by a boron atom whilst retaining the original dimensions of the structure overall is broadly equivalent to introducing a tensile strain of about 5 x 10⁻⁴. Furthermore this strain *may* prove to be unstable due to potential future plastic flow. This topic will be returned to later in the paper.

The drive whereby these microresonators are excited is usually opto-thermal and the driving mechanisms are discussed in more detail in the next section. However, for an opto-thermal drive to be successful it is fundamentally clear that the beam must absorb the optical energy which is somehow imposed upon it, typically by illuminating it at the centre as indicated in Figure 2. Optical fibres are best suited to use in the near infra-red and silicon is a poor absorber at these frequencies. In particular for wavelengths longer than that corresponding to the band gap (1.1μm) silicon is effectively transparent. Consequently most silicon microresonator transducers designed for optical excitation require some form of absorbing layer to provide the opto-thermal drive and usually this is provided by depositing a metal such as aluminium or preferably chromium on the surface. The oscillator is then a composite structure (see Fig.3) though the metal layer is thin compared to the silicon underneath it. The mechanical properties of silicon (see Table 1) dominate the behaviour of the overall resonator but the metal layer, thanks to differential thermal expansion can apply a (sometimes beneficial) thermally tuned stress to the composite beam.

As a final observation on silicon resonators, Table 1 summarises the mechanical properties of silicon compared to many common structural materials. The only significant differences lie in linear coefficients of thermal expansion. This difference is very important in determining the necessary packaging procedures required to ensure stable operation of a particular transducer over a wide temperature range.

The basic principles are then as follows. A silicon microresonator is etched into a substrate material using the process outlined above exploiting a boron etch stop or other similar techniques. The desired measurand (usually but not always pressure) is then made to alter the tension within the resultant resonator. The geometry of the resonator determines its resonant frequency, the resonant frequency spectrum and the behaviour of this spectrum under variations in tension. Detailed design procedure are complex and often need to resort to finite element analysis for fine tuning of the results but the principles are always in line with those described in the preceding paragraphs.

3. OPTICS FOR MICROMACHINED SENSORS

The optical system has two principal functions. First of all it must transmit power to the sensor head and translate this power as efficiently as possible into mechanical movement. In order for the micromachined sensor to realise its full potential this translation process should not include any electronic stages. The second function of the optical system is to detect the consequent motion of the resonator element. For typical resonators this motion has a peak to peak amplitude of around 100nm. The detection system must obviously be capable of monitoring these small displacements with a good signal to noise ratio to facilitate the accurate measurement of the resonant frequency at the receiver.

In general these two functions are undertaken by two different sources which are wavelength or time multiplexed on to an optical fibre linking the control area to the sensor. Figure 4 illustrates schematically the system format.

3.1 Optical drive mechanisms

The use of a direct optical to mechanical conversion mechanism eliminates an optical - electronic - mechanical sequence and the presence of electronics within the sensor head would undoubtedly compromise its high temperature performance. However in circumstances which do not require the high temperature capability this indirect drive mechanism can be extremely effective and is probably the most efficient conversion technique available despite the intermediate electronic stage. Indeed early versions of optically interrogated and excited resonator transducers were based upon exactly this principle[4]. Since the optics can be added as a bolt-on extra to a conventional transducer these systems are relatively simple to realise (see Fig. 5) and have already been developed into commercial product.

There are three "direct" mechanisms whereby light may be transformed into mechanical movement. These are:

- radiation pressure

- conversion of optical intensity into electronic strain in semiconductors

- opto - thermal techniques.

Of these only the first is truly a direct mechanism. The value of the radiation pressure is given by:

$$F_{rad} \leq \frac{2P_{opt}}{c} \qquad\qquad 6$$

The maximum optical power available in these systems is of the order of 1mW so the maximum radiation pressure (strictly of course it should be called a radiation force) is of the order of 10pN. The effective stiffness of a clamped-clamped bridge structure is given by[5]:

$$\beta_{cc} = \frac{\delta_m}{F_m} = \frac{L^3}{96EI} \qquad\qquad 7$$

where δ_m is the deflection at the centre of the beam as the result of the application of the force F_m at the centre of the beam and I is the moment of inertia given by:

$$I = \frac{wt^3}{12} \qquad\qquad 8$$

Substituting values into equations 7 and 8 for the most flexible of transducer systems consisting of a beam 1mm in length and 10 microns wide and 2 microns thick gives a maximum deflection of the order of 0.1nm. Whilst this is in principle detectable the actual deflection values observed are found to be orders of magnitude higher than this so that radiation pressure drive is not considered to be an important contributory phenomena.

Electronic strain is introduced in semiconductors when a non-equilibrium carrier distribution is introduced into the crystal. This clearly happens when light is absorbed to produce a photocurrent. The strain is given by:

$$\varepsilon_{el} = \frac{1}{3} \frac{dE_g}{dP} \Delta n \qquad\qquad 9$$

where E_g is a band gap of the semiconductor, P is pressure applied to the semiconductor and Δn is the non-equilibrium local carrier distribution. In silicon the order of magnitude of this effect has been shown to be comparable to that of photothermal phenomena in high quality bulk silicon. However the drive mechanisms in microresonators operate upon heavily boron doped thin silicon layers in which the carrier lifetime is extremely small and is dominated by surface recombination. In silicon the rate of change of band gap with pressure is negative and the electronic strain is therefore of opposite sign to any photothermal phenomena. The circumstantial evidence is the electronic strain in the material which forms the resonator element is negligible due to the fast recombination time. However definitive experiments to unambiguously demonstrate that this is indeed the case have not yet been reported in the literature and since the photothermal and electronic drive mechanisms always co-exist separation of the two promises to be extremely difficult within any viable resonator element.

All the evidence indicates that the photothermal drive dominates the optical to mechanical conversion process. The driving force is usually a sinusoidally modulated semiconductor laser which may be characterised by (Fig. 6) mean power I_m and a peak modulation intensity I_f. Of this optical power some fraction will be reflected from the surface, some will be absorbed within the resonator and some will be transmitted through it. For the uncoated resonator with a typical thickness of 2µm approximately 30% is reflected, 10% absorbed in the resonator and the remainder transmitted (at a wavelength of 820nm). The addition of a thin absorbing metallic layer changes these ratios to approximately 50% reflected and 50% absorbed within the metal driving layer. For coated structures these numbers depend upon the coating material and also upon its thickness but since the absorption depth in metals is at most tens of nanometres a very efficient absorption can be assured.

In order to understand the thermal driving process it is useful to consider the effects of the mean power I_m and the modulation I_f separately. The static component gives rise to a temperature distribution which is a simple linear function from the centre (where the driving point is assumed to be) to the ends of the beam. The temperature rise Δt is given by:

$$\Delta T = \eta I_m R_\theta \qquad\qquad 10$$

where η is the fraction of incident light which is converted to heat and R_θ is the thermal resistance of the beam and which when viewed from the centre consists of two equal sections in parallel each of length $L/2$:

$$R_\theta = \frac{L}{4wt\,\sigma_{th}} \qquad\qquad 11$$

where σ_{th} is the thermal conductivity. The maximum typical value of R_θ is again for a 1mm long 10µm wide and 2µm thick beam and is of the order of 8×10^{4}°C/watt. Typical beams are of the order of one-third of the total length and twice the width giving a typical thermal resistance of 10,000°C/watt. Of the incident optical power rarely more than hundreds of microwatts are converted to heat so that the maximum temperature rise in the centre of the beam is typically 1°C. This is sufficient to cause minor buckling of the beam structure. In order to estimate this we can assume that the ends are clamped and that the buckle structure assumes a sinusoidal shape. We then have a displacement A from the equilibrium position of the beam which is given by approximately:

$$A \quad \simeq (2\pi\alpha\Delta T)^{\frac{1}{2}}L$$
$$\simeq 3 \times 10^{-3} \, L \qquad\qquad 12$$

so that the static thermal contribution displaces the centre of the beam upwards by or the order of 1μm. The dynamic contribution is that which gives rise to the detected motion of the resonator.

In order to understand the dynamic behaviour of a beam under thermal excitation we need to first of all derive the dynamic version of equation 10 and then determine the consequent dynamic strain distribution within the beam. The key is to extract a high frequency version of the thermal impedance and in order to do this it is simplest (at least for the electrical engineer) to view the beam as a thermal RC transmission line whose characteristic impedance $R_{\theta\omega}$ is given by:

$$R_{\theta\omega} = (\frac{R}{2\omega C})^{\frac{1}{2}} (1 - j) \qquad\qquad 13$$

where R and Z are the thermal resistance and thermal capacitance per unit length of the beam given as:

$$R = \frac{1}{\sigma_{th}wt} \qquad\qquad 14$$

$$C = s\rho wt$$

where s is the specific heat ρ is the density and σ_{th} the conductivity of the material comprising the beam.

The propagation constant γ for the thermal wave travelling along the transmission line is:

$$\gamma = (j\omega RC)^{\frac{1}{2}} \qquad\qquad 15$$

which gives us a penetration depth δ (usually termed the thermal diffusion depth by thermodynamists and the skin depth by electrical engineers) of:

$$\delta = (\frac{2\sigma_{th}}{\omega s\rho})^{\frac{1}{2}} \qquad\qquad 16$$

This corresponds to a highly dispersive, rapidly attenuated wave and provided that δ is greater than L the effect of the thermal "reflectived wave" from the short circuit at the substrate end of the beam can be neglected. The values of δ are plotted in Figure 8 for silicon, aluminium and chromium from 10kHz to 1MHz. This figure graphically illustrates that for most, if not all, circumstances of interest the travelling thermodynamic wave can be considered to be propagated into a matched load.

We could also assign a group velocity, the v_{th} to the thermal wave of:

$$v_{th} = \frac{\partial \omega}{\partial \beta} = (\frac{2\omega}{RC})^{\frac{1}{2}} = (\frac{2\omega\sigma_{th}}{s\,\rho})^{\frac{1}{2}} = \omega\delta \qquad\qquad 17$$

This is also plotted on Figure 8. These parameters characterise the thermal wave which is generated by the dynamic thermal source at the centre of the beam structure. The peak temperature variation at the centre of the beam is simply:

$$\Delta T_{\omega} = I_{\omega} R_{\theta,\omega} \qquad\qquad 18$$

and from equations 13 and 18 we can see that ΔT_{ω} decreases as the square root of ω and also depends upon I_{ω} which is a measure of the modulation depth of the driving optical beam. The higher the value of this the greater the drive. For example a 100% modulated square wave drive will be more effective by a ratio of $4/\pi$ than a 100% modulation depth sinusoidal signal. Note that provided that δ is less than L, ΔT_{ω} is independent of the length of the beam structure. The value of ΔT_{ω} as a function of frequency for silicon beam of width 10µm and thickness 2µm as a function of frequency between 10kHz and 1MHz is shown in Figure 9 for 1mW drive.

We can consider a thermal wave propagating along the beam originating in the centre and given by:

$$\Delta T(z,t') = \Delta T_{\omega} e^{-j(\gamma\,z - \omega t')} \qquad\qquad 19$$

where t' is the time co-ordinate, z is the co-ordinate along the beam originating at the centre and γ is the propagation constant given in equation 15. This will in turn generate a thermal strain wave given simply by:

$$\varepsilon(z,t') = \alpha\Delta T(z,t') \qquad\qquad 20$$

This travelling strain wave is then converted into the mechanical movement of the beam at the frequency ω. Equation 20 assumes that the ends of the beams are clamped so that the temperature changes do indeed convert into a directly equivalent strain change. It also assumes that the resultant strain distribution in the beam exactly matches that in the equation. This is manifestly untrue since the beam will assume the appropriate mode shape. However, equation 20 does give a guide as to the *maximum* mechanical strain which can take place.

In order to ascertain the strain energy E_{ε} input to the beam by the thermal wave, we need to evaluate the integral :

$$E_\varepsilon(t') = \frac{Ewt}{2} \int_{-\infty}^{\infty} \varepsilon(z,t')\, \varepsilon(z,t')\, dz \qquad\qquad 21$$

where the integral over z is taken over the entire length of the beam which, for the case where $L > \Delta$ may be approximated by an integral from $-\infty$ to ∞.

In order to ascertain the power flow into the beam at a particular instant we need to take the time derivative of equation 21. Integrating equation 21 gives :

$$E(t^2) = \frac{\alpha^2 \Delta T_\omega^2 Ewt}{4} \left[\; 1 + 0.5 e^{j2\omega t'}(1 + j) \;\right] \qquad\qquad 22$$

giving the *total* energy storage to beam at a particular time t' under the assumption δ is less than L. Taking the derivative of this equation gives the power flow at a particular time t' as :

$$P(t') = \frac{\alpha^2 \Delta T_\omega^2 Ewt\delta\omega}{4} \cdot je^{2j\omega t'} \left[\, 1 + j \,\right] \qquad\qquad 23$$

After some substitution we find that the power flow is given as :

$$P(t') = I_\omega^2 \cdot \alpha^2 E \left[\; \frac{2}{\sigma_{th} s^3 \rho^3} \;\right]^{\frac{1}{2}} \cdot \frac{1}{wt} \cdot \frac{je^{2j\,\omega t}(1\text{-}j)}{2\sqrt{\omega}} \qquad\qquad 24$$

From this equation we can deduce the following :

- there is a complex materials related influence on the achievable power flow

- the structural parameter which is most important is the cross-sectional area

- the total power flow is proportional to the square of the input power and consequently ....

- the efficiency of the power conversion process increases linearly with the drive amplitude

- the power transfer capability of the system decreases but relatively slowly (as a square root) as the frequency of the drive is increased.

Equation 24 is the fundamental power transfer from the thermal input drive into mechanical energy. Other mechanisms come into play of which the most important is the effect of the mode shape of the mechanical resonance which is to be excited within the material. The drive has a particular set of characteristics given essentially by equation 20 and the efficiency of this drive in converting to the mode of interest can be simply derived in principle by using a mode overlap integral of the type :

$$I_{overlap} = \int_{-\infty}^{\infty} \varepsilon_m(z,t')\varepsilon(z,t')dz \qquad\qquad 25$$

where ε_m is the strain distribution corresponding to the mode being excited. The thermal strain distribution is symmetrical about the excitation point so the most efficient excitation into mechanical oscillation is for modes which are symmetrical about this point. Furthermore, ideally the modal wavelength should match to the thermal excitation spatial function, so should be of the order of δ at the frequency of interest.

As a final observation, the drive is in essence via a heat engine with a relatively small temperature differential between heat source and heat sink of ΔT_ω. In this case the efficiency is simply limited to the value η given as :

$$\eta \leq \frac{\Delta T_\omega}{T} \qquad\qquad 26$$

The values which are achievable in a practical environment η will always be less than 0.1% though the apparent efficiency can be increased through the effects of energy storage through the Q factor. This efficiency limitation compounds the earlier observation that there are intrinsically much more effective techniques involving an electronic interface whereby the excitation process may be put in place. However, the sensitivity of optical detection enables the reliable interpretation of the relatively small displacements which can be achieved at this efficiency level. The applications need for electronically passive sensor elements for use in specific environments has stimulated the continuation of research into this class of device.

3.2 Detection processes

The vibrational amplitudes obtained in silicon micromachined resonator sensors are typically in the range of 10-100nm. Consequently optical detection is best performed interferometrically. The type of detection system to be used depends upon whether or not quantitative measurements of the vibrational amplitudes are required and whether or not self oscillation induced through the optical arrangement is expected.

The interferometric transfer function is the well known "cosine squared" dependence of intensity against path difference between the two arms of the interferometer. The signal which is obtained for a particular ac modulation on the bias differential between the paths is simply obtained through the slope of the interferometer response curve at that particular bias level. When the differential is zero there is no signal at the fundamental frequency of the dynamic displacement and a (relatively) small signal obtained at the second harmonic. Consequently any interferometric detection system must recognise this dilemma and incorporate the necessary optical and/or electronic processing to enable reliable displacement detection no matter where the interferometer is statically biased. The alternative is, of course, to lock the interferometer on a particular value of the static bias which is exactly one or other quadrature phase point and to maintain this bias value to the necessary tolerance (typically 50nm). Since the applications interest is primarily for pressure transducers in which the vibrating element is etched into the back of a diaphragm and the diaphragm moves over significant distances (microns), this second approach is usually impractical. Another option is, of course, to ensure that the displacement amplitude is sufficiently large for shadow mask approaches to give intensity modulation of an optical beam. This again is generally undesirable since it requires relatively large excitation powers and also necessitates relatively large displacements over which the mechanical oscillations may well become non-linear distorting the apparent value of the oscillation frequency.

There are two established approaches to the realisation of path length independent interferometric detection systems. There are :

- the use of heterodyne interferometry

- the use of some form of effective quadrature differential between two real or virtual reference paths in the optical system followed by electronic combining.

A third approach has been to use 3 x 3 couplers and exploit the "three phase" outputs and again use electronic combining. In effect this is a special case of the second of the above options but with 120° rather than 90° offsets. This approach has been used to good effect in simple, low cost fibre optic gyroscopes but the 3 x 3 coupler does implicitly require that the entire fibre optic system be single mode. The environment within which a pressure transducer is likely to operate could well necessitate the use of multimode fibres to enable more relaxed mechanical tolerancing of peripheral piece parts.

There are many approaches to interferometric detection which can be utilised. Two which we have found particularly useful are shown in Figures 10 and 11. The arrangement shown in Figure 10 is a modified heterodyne Mach Zehnder operating at an intermediate frequency of typically 80MHz. The vibrating element introduces phase modulation at the vibrational frequency on the signal arm of the interferometer and this in turn is translated directly to the intermediate frequency. By measuring the relative amplitudes of the side bands, the amplitude of the vibration is very easily derived. This approach has proved to be invaluable in obtaining quantitative information on the displacement characteristics of these devices and it obviously fulfils the requirements of sensitivity and bias tolerance outlined earlier.

The second approach shown in Figure 11 is designed for use in a practical environment. The principle is simply to use white light interferometry which can, of course, be implemented via multimode fibre but to build into the reference interferometer two independent paths which are phase offset by 90°. A particularly simple technique for realising this phase offset is to define two polarisation axes (via the polarising beam splitter in the diagram) and to use a 1/8th wave plate (or suitable multiple thereof) to provide the necessary offset[6]. The relatively standard differentiate and cross multiply electronics can then be used to extract the required signal.

In their present form neither of these techniques is readily adapted for self oscillation optical systems of the type shown in Figure 12. In a self oscillation system a signal optical source is used for both excitation and detection whereas for the system shown in Figures 10 and 11 a separate source is required for excitation and a feedback loop between the source and detector provides electronic feedback. Single optical source systems have been realised. Most frequency by using an appropriately biased Fabry Perot interferometer with a feedback mechanism. However, the mechanical tolerances required here are even stricter than those required for maintaining quadrature bias in a conventional two beam interferometer. The Fabry Perot approach, whilst of interest, is dogged by the practical issues implicit in building the necessary interferometric system of which the vibrating element itself must form one end and the - effectively - transducer package the other[7].

Perhaps a more practical approach is that shown in Figure 12(b) which effectively simplifies the optics for a dual source system by time gating a single source to act in both excitation and detection roles. An impulse from the source starts the oscillation of the resonator which can be maintained if the duty cycle is of the order of Q^{-1}. However, instead of switching the source completely off during the "space" part of the cycle, a low power is maintained in order tp observe the movement of the vibrating element[8]. Whilst the loop has not been electronically closed for such systems, there is the possibility for a frequency dividing phase lock loop approach to maintain stable operation.

3.3 Discussion

The opto-thermal drive mechanisms, whilst inefficient, is certainly more than adequate to provide detectable vibrational amplitudes. Much has still to be learnt about the detail of this mechanism and in particular about the effects of the spatial matching between thermal and acoustic modes. In practice power levels of less than 1mW will always provide more than adequate displacement amplitudes and oscillations can be induced through either single or multimode fibre depending upon the detailed structures of an individual mechanical element.

Detection techniques are relatively simple though do require path length tolerant interferometric systems. Current approaches largely involve separate detection and excitation sources though a number of single source systems have been demonstrated. The need for remote interrogation through often the very long lengths of optical fibres does impose the necessity for an interferometer at the sensor head. The white light approach appears to be a particularly attractive solution to this problem since it can be adapted to provide reference path independent operation to work with single or multimode fibres and to operate over extremely long lengths.

4. EXPERIMENTAL OBSERVATIONS

This section presents the principal features of a considerable portfolio of experimental data amassed by several research teams in the US and Europe. Virtually all the experimental results are concerned with vibrating beam structures which are refinements on the simple concept shown in Fig. 2. The remainder of this section is divided into three parts. In the first we consider the basic properties of the beam structures themselves, then we examine some examples of the use of these beam structures as transducers and finally we make some observations on current experimental trends and topics for further research.

4.1 Basic properties of vibrating beams

The results of measurements on a wide range of vibrating element devices are shown in Fig. 13. These resonant frequencies are all well above the values for an unstressed silicon beam so that the effects of tension due to boron doping are very significant. By plotting f^2L^2 against $1/L^2$ and taking a least squares fit we can from these results derive the mean strain level induced by the boron doping and we find it to be 6.8×10^{-4} which agrees well with the intuitive reasoning presented in section 3.

It is natural to assume that annealing could stabilise and reduce the value of the dopant induced strain. Table 2 shows the results of annealing a set of structures at 1100°C in nitrogen for 50 minutes. There is obviously some stress relief since the resonant frequencies have decreased though we found that further annealing at high temperatures made no measurable difference to the resonant frequencies.

We have already alluded to the fact that a coating is necessary in order to produce the thermal drive. The coating using aluminium or chromium also has an effect on the resonant frequency behaviour of the device. In particular the coating (see also Table 2) reduces the resonant frequency further and also has an influence on the Q factor. The effects for aluminium on silicon are shown in Fig. 14. Aluminium is inherently a lossy material and considerably improved results can be obtained by using a less lossy acoustic material such as chromium.

These results can also be interpreted to arrive at effective values for the materials properties of the beam and the coating in thin film form. For silicon the Young's Modulus of the beam is found to 12.2×10^{10} N.m^{-2} a little below the bulk material value. The reduction is put down to the presence of the boron atoms within the lattice. Surprisingly the agreement between the measured value of the Young's Modulus of 75nm of aluminium ($6.93 \pm 0.07 \times 10^{10}$ N/m^2) and the quoted value of 6.9×10^{10} N.m^{-2} is almost perfect so that the deposited the thin film behaves equal in the bulk material.

The principal global conclusions of these measurements on the basic properties of beams are:

- the resonance properties of the beam structures can be predicted reasonably accurately using the mechanical properties of the bulk materials provided that due recognition is made of any in-built stresses

- the stresses induced by boron doping used as an etch stop are significant and can dominate the resonance behaviour of the structure. These stresses can be reduced and stabilised by annealing processes

- the Q factors of the resulting resonators are strongly influenced by both the presence or otherwise of air around the resonator and the coatings used to provide the optothermal drive. At moderate vacuums Q factors of a few thousand can be obtained and these can be increased to tens of thousands under higher vacuum even for very simple vibrating elements based upon the structure shown in Fig.2. Structures designed specifically for low losses can achieve Q factors in excess of 50,000

- the simple beam structure has an overtone spectrum which is extremely simple and which enables straightforward discrimination between excited modes over the range of resonator frequency values through which a transducer may be exercised. The more complex the structure the more lively that excitable resonances appear close to each other in frequency space. In particular structures which have nominally degenerate modes should be totally avoided (for example, structures with nominally square or circular symmetry).

4.2 Transducer Measurements

Many measurands can be interfaced to the basic vibrating bridge via some form of strain conversion interface. The simplest interface is probably that between the vibrating element and temperature changes which can be introduced by, for example, mounting the vibrating element on material (for example, iron) which has a large differential coefficient of thermal expansion with silicon. A similar effect may be obtained by providing a relatively thick metal coating on the beam substrate and mounting structure. Differential thermal expansion between the silicon substrate and the layer deposited on the beam also induces a temperature dependent stress. By using different combinations of these to basic techniques the temperature coefficient of the resonant frequency of the bridge can be adjusted over a range of $\pm 0.5\%/°C$. Typical experimental results for a bridge mounted upon an aluminium substrate are shown in Fig. 15.

The concept is however probably more useful as a pressure transducer and indeed it is in this area in which most of the interest has been stimulated. In pressure transducers the vibrating beam is subjected to variable tension in response to the bending of a diaphragm upon which the beam is mounted. The diaphragm is usually fabricated from silicon and in its simplest form a bridge such as that shown in Fig. 2 is etched on the reverse side of the diaphragm itself. However it is usually desirable to incorporate some scale factor gain so that in most practical versions of these devices the bridge is mounted on pillars to distance it from the reverse side of the diaphragm and therefore increase the strain applied to the bridge for a given flecture of the diaphragm (see Fig. 16 which shows the principles of one of our own devices). The behaviour of this particular device giving the frequency change as a function of pressure is shown in Fig. 17.

This structure is based upon a boron doped beam and also requires illumination from an axis which is vertical to the plane of the diaphragm. An alternative and more convenient approach which requires illumination from the side of the diaphragm is shown in Fig. 18. This geometry has the additional advantage that it uses unstressed silicon as the beam so the resonance behaviour depends entirely upon the stiffness characteristics of the silicon and not upon any doping induced stresses. Fig. 19 shows the results obtained from this transducer and also indicates the temperature coefficient of the resonant frequency which is approximately 30 parts per million/°C.

Several other organisations have also examined micromachined structures. One of the most important has focused upon down hole monitoring applications for use in geophysical exploration. In this particular demonstration a transducer has been operated at the end of a 25km length of optical fibre and has shown excellent pressure monitoring performance through temperature ranges in excess of 200°C.

4.3   Experimental results - discussion

The principal area of interest lies currently in using vibrating element devices as pressure transducers, in particular for applications in harsh environments. The principal difficulties lie in the temperature coefficients of these devices and by implication (see Fig. 15) in the engineering of the mounting structures in which the devices are operated. The approach taken to achieve the results in Fig. 19 was essentially to use an all silicon packaging technique with a final encapsulation physically a long distance from the actual measurement head and coupled through stress relieving silicon substructures.

The intrinsic temperature coefficient of a silicon based resonator is essentially dominated by the rate of change of the Young's Modulus with temperature (see equations 4 and 5). For silicon the value of this parameter is in the region of -55 parts per million/°C so that the intrinsic temperature coefficient of resonant frequency is -27 parts per million/°C. The value obtained in the results in Fig. 19 is only a little higher than this indicating that extremely good stress relieving has been incorporated into the structure.

To achieve even better thermal compensation some form of temperature tuned stress should be introduced. In principle this can be achieved by putting a suitable coating on the vibrating element and the results in Fig. 20 illustrate that with the appropriate coating temperature coefficient of better than 5 parts per million/°C are in principle achievable. Given the availability of a suitably stress relieved overall package these results demonstrate that extremely stable pressure measurement should be realisable using the optically excited vibrating element silicon micromachined system.

5. CONCLUSIONS

The first results indicating the potential for optical excited vibrating element silicon as a transducer structure were published in 1985[10] though optothermal drive of small mechanical structures has been known for considerably longer. Since then considerable progress has been made towards the realisation of usable practical transducer elements and a substantial enhanced appreciation of the applications potential of these devices has emerged. In particular their potential in high temperature environments is unmatched by any other transducer technology. In the short term it will be these applications (for example, gas turbine engine monitoring) which will drive the technology. However there is considerable potential for spin off into lower cost and less demanding arenas such as, for example, medical instrumentation where there the batch process capability of silicon micromachining will come to the fore.

Additionally there is considerable as yet untapped potential for the use of this basic technology to monitor a much wider range of measurands. The basic structure can be configured to be sensitive to either pressure or temperature or indeed some combination of the two by suitable design or the multiple layer vibrating element and its mounting structure. Temperature and strain fields can be created through measurand transformers from most physical and chemical parameters thus for example the technology can be adapted to produce magnetic field measurements, chemical parameter measurements, acoustic field measurements, flow rates etc etc. The current level of industrial interest in vibrating element systems promises significant future developments. Whilst the detailed engineering still has a long way to go the basic principles whereby the engineering problems may be solved have all been demonstrated and this coupled to the ability presented by vibrating silicon to successfully address otherwise intractable measurement problems will ensure the realisation of production devices within the imminent future.

6. ACKNOWLEDGEMENTS

The results presented in this paper reflect the contributions of several research students and postdoctoral fellows at Strathclyde over the past few years including Katherine Thornton, Li-Ming Zhang, Douglas Walsh, John Nixon, John Chalmers and my colleague Deepak Uttamchandani. I would also like to thank Bruce Hockaday of UTRC for providing access to the results in Figures 18 and 19 and our research sponsors including the UK SERC, OSCA, Foxboro Company, Boeing, STC (now BNR Europe), Honeywell, Schlumberger and TransInstruments for their support. Our work has also benefited from discussions with many people outwith the University including Tony Gilby, Roger Jones, Phil Parsons, Peter Fry and many others.

7. REFERENCES

1.   R M Langden "Resonator sensors - a review" J Phys E, Sci Instrum, 18, 103-115 (1985)
2.   K E Petersen "Silicon as a mechanical material", Proc IEEE 70, 5, p420-457, (1982)
3.   M Rossi "Acoustics and electro-acoustics" Artech House Norwood Mass 1988 (Chapter 5)
4.   R C Spooncer, B E Jones and G S Philp "Hybrid and resonant sensors and systems with optical fibre links", Journal IERE, 58, 5 p585-591, 1988
5.   J M Gere and S P Timoshenko "Mechanics of materials", (2nd Ed) Brooks/Cole Engineering 1984.
6.   D Walsh PhD Thesis University of Strathclyde, Glasgow, Scotland 1992
7.   L M Zhang, PhD Thesis University of Strathclyde, Glasgow, Scotland 1990

8. L M Zhang, D Uttamchandani, B Culshaw "Transient excitation of silicon microresonator" Electronics Letters, 25, p149-150, 1989

9. B Hockaday, Late Submission to OFS 7 Sydney Australia 1990 and private communication 1992

10. S Venkatesh and B Culshaw, "Optically activated vibration in a micromachined silicon structure", Electronics Letters 21, 315-317, 1985.

TABLE 1

Properties of Silicon, Chromium, Aluminium and High Strength Steel

Material	Density kgm.m^{-3}	Thermal Conductivity W.m^{-1}	Thermal Expansion 10^{-6} °C^{-1}	Specific Heat J kgm^{-1} °C^{-1}	Youngs Modulus x10^{10}Nm^{-2}	Yield Strength x10^{8}Nm^{-2}
Silicon	2.23	156	2.5	714	11	70
Chromium	7.2	91	6	462	25	-
Aluminium	2.7	237	24	903	6.9	1.7
High Strength Steel	7.9	970	12	-	21	42

TABLE 2

Beam Length (mm)	RESONANT FREQUENCY (kHz)		
	p$^+$-Si	P$^+$-Si (annealed)	p$^+$-Si + 75nm Al
1.28	84.220	82.425	81.351
0.86	126.296	124.290	121.761
0.72	149.483	148.428	146.046
0.58	190.431	187.537	184.038
0.45	253.633	249.173	244.652
0.28	432.994	425.982	420.970
0.24	524.633	518.602	513.555

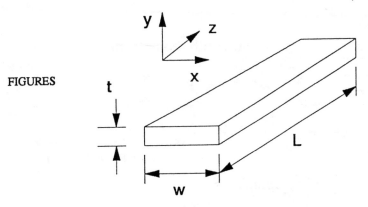

FIGURES

Figure 1 Basic geometry of flat rectangular beam resonator

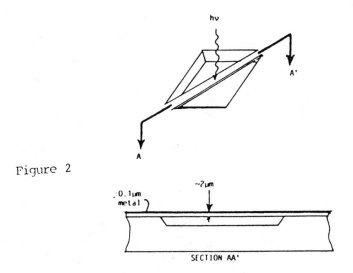

Figure 2

Excitation process and structure for a typical experimental vibrating bridge.

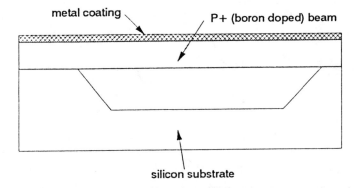

Figure 3 Composite silicon microresonator structure

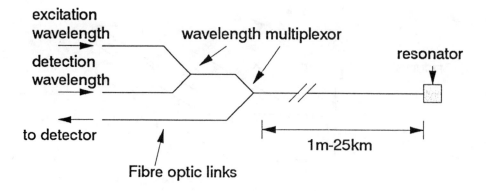

excitation
wavelength

detection
wavelength

wavelength multiplexor

resonator

to detector

Fibre optic links

1m-25km

Figure 4 Dual wavelength optical system schematic

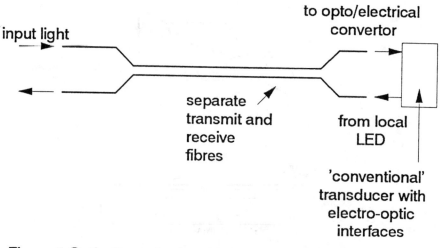

to opto/electrical
convertor

input light

separate
transmit and
receive
fibres

from local
LED

'conventional'
transducer with
electro-optic
interfaces

Figure 5 Optically excited and interrogated vibrating
transducer system with electro-optic drive

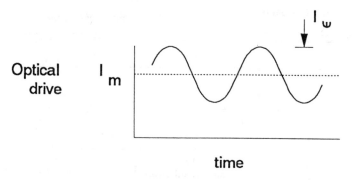

I_ω

Optical
drive

I_m

time

Figure 6 Driving function for opto-thermal conversion

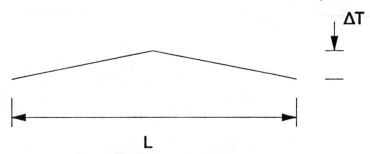

Figure 7 Quasi-static temperature distribution during opto-thermal drive

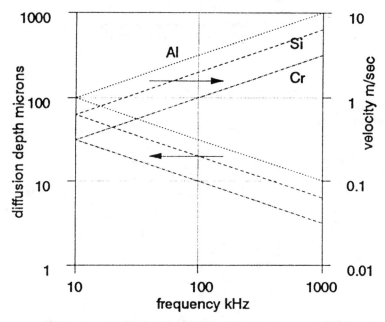

Figure 8: Diffusion depths and thermal velocities

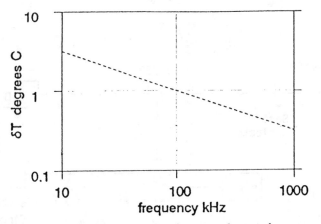

Figure 9 Dynamic temperature changes for peak power modulation of 1mW onto Si beam 2μm x 12μm with L<< diffusion depth

Figure 10

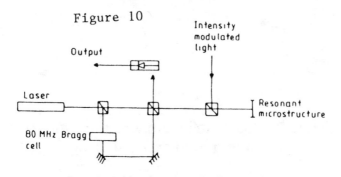

Combined interferometer displacement
and intensity modulated excitation

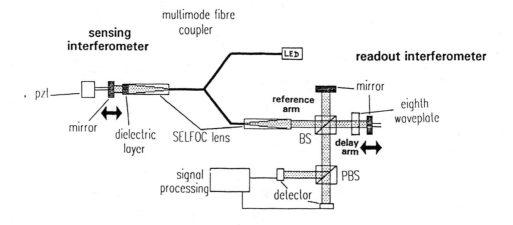

FIGURE 11: Passively compensated white light interferometric system.

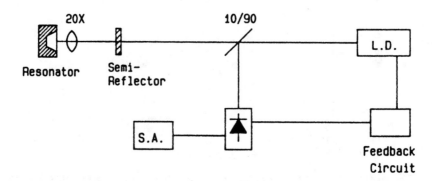

Figure 12(a)   Arrangement for the stabilisation of self-oscillations

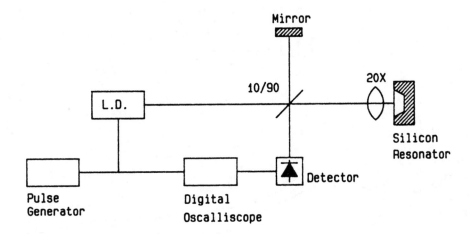

Figure 12(b)   Arrangement for transient excitation

FIGURE 13:. Least mean square plot of $f_r^2 L^4$ against $1/L^2$ for an array of p^+-Si beams with t=2.1μm and w=60μm.

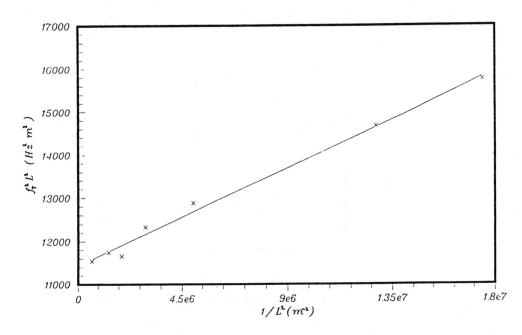

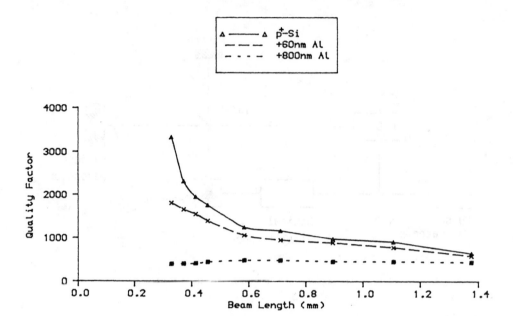

Fig. 14: **Quality factor at 4mbar for an array of p⁺-Si beams with various thicknesses of aluminium; t=2.1μm and w=24μm.**

Figure 15

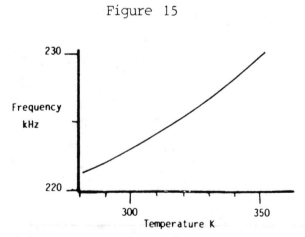

typical resonant frequency versus temperature behavior for a
metal-coated bridge structure.

Figure 16: Micrograph of pressure
transducer showing a resonator
connected to a 1mm square
diaphragm

Figure 17

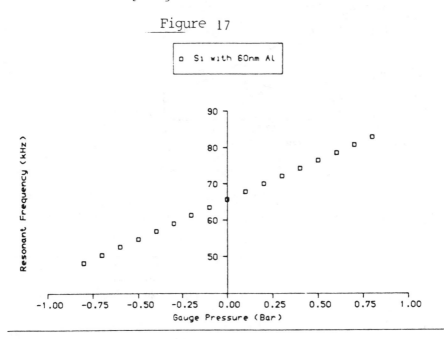

Resonant Frequency against pressure

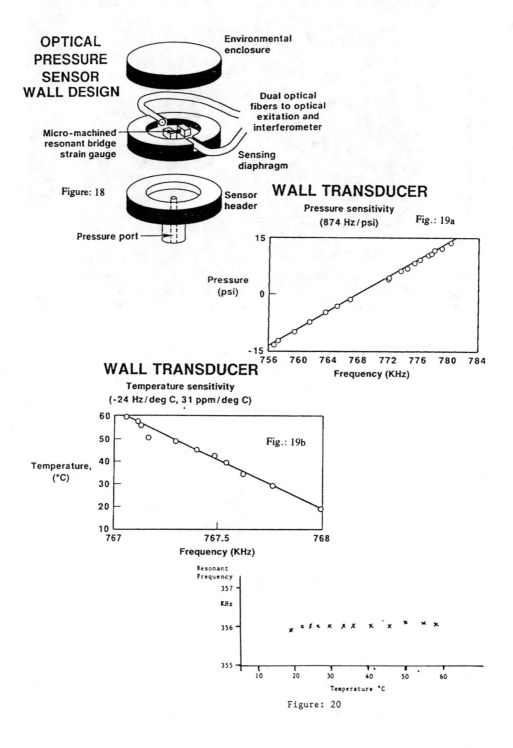

OPTICAL PRESSURE SENSOR WALL DESIGN

Environmental enclosure

Dual optical fibers to optical exitation and interferometer

Micro-machined resonant bridge strain gauge

Sensing diaphragm

Figure: 18

Sensor header

Pressure port

WALL TRANSDUCER

Pressure sensitivity (874 Hz / psi)

Fig.: 19a

Pressure (psi)

Frequency (KHz)

WALL TRANSDUCER

Temperature sensitivity (-24 Hz / deg C, 31 ppm / deg C)

Fig.: 19b

Temperature, (°C)

Frequency (KHz)

Resonant Frequency

KHz

Temperature °C

Figure: 20

Grating and polarimetric-based fiber sensors

W. B. Spillman, Jr.

Physics Department, University of Vermont, Burlington, Vermont 05405

ABSTRACT

Since the initial demonstrations of various fiber optic sensing techniques in the laboratory, much progress has been made in creating practical sensors for use in real world applications. In particular, sensors utilizing the properties of optical gratings and sensors exploiting polarization effects have been intensively investigated. Both multimode and single mode fiber sensors have utilized these transduction techniques. In this paper, a brief review of grating and polarization optics will be provided followed by a more detailed analysis of a number of specific multimode and single mode fiber optic sensors. Included among these will be hydrophones, pressure sensors, strain sensors, vibration sensors, linear position sensors and speed/torque sensors. The practical application of these sensors will also be discussed.

1. INTRODUCTION

Although the field of fiber optic sensing has been in existence for a considerable period of time for applications such as internal viewing of the human body (endoscopes), it has only begun to display a wider potential since the development of low loss optical fiber for telecommunication purposes. At the present time, a very large number of different fiber optic sensing mechanisms have been proposed, demonstrated in the laboratory and reported [1-4]. A number of these sensors have been developed to the point at which they can be used for practical applications. They no longer represent just laboratory curiosities. Two classes of fiber optic transduction mechanisms have shown themselves to be of general utility and have been used to sense a number of different parameters. These transduction mechanisms have been demonstrated in both intrinisic sensors (the state of the optical signal is modulated within the fiber) and extrinsic sensors (the light is removed from the fiber, acted upon by the transduction mechanism and then reinjected into the fiber). The first transduction mechanism modulates the state of an optical signal in various ways via a grating structure while the second modulates the polarization state of an optical signal. Discussion of the details of a number of sensors based on these two transduction mechanisms will be the subject of this paper. A theoretical background will first be provided in which grating and polarimetric (polarization) optics will be briefly discussed. Following that, a number of sensors based on grating modulation will be analyzed including hydrophones [5], pressure sensors [6], rotary [7] and linear [8,9] position sensors and distributed strain sensors [10,11]. A number of sensors based on polarization state modulation will then be discussed including rotary position sensors [12], pressure sensors [13-15], accelerometers [16], hydrophones [17], vibration sensors [18], proximity sensors [19], speed/torque sensors [20,21] and magnetic field sensors [22]. Finally, the practical application of these sensors and the future of fiber optic sensing will be covered.

2. THEORETICAL BACKGROUND

Although a rigorous discussion of the optics and other physical principals of fiber optic sensors is beyond the scope of this paper, some general formalism needs to be introduced to provide a common framework for considering the various sensing techniques. In particular, the concept of a transfer function along with grating and polarization optics have considerable importance in understanding the operation of the various sensors and their limitations.

2.1 Transfer function

In general terms, any transducer, optical or otherwise, performs in a mathematical sense as a transfer function, T. This function operates upon some input state, $\{I_0\}$, and modulates it based upon some parameter of interest, X. To a lesser extent, moduation is also produced by changes in the environment surrounding the transducer as manifested through changes in the environmental parameters, $\{E_i\}$. A schematic diagram of a fiber optic sensor with its transfer function is shown in Figure 1.

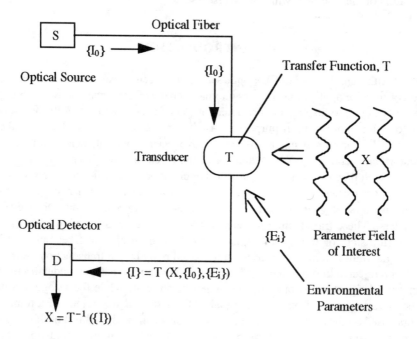

Figure 1

As shown in Figure 1, light from an optical source, S, is injected into an optical fiber. The light in the optical fiber is described by a state, $\{I_0\}$. The optical state includes intensity, polarization, wavelength and other factors for every mode traveling in the fiber. The light then passes through the transducer with its transfer function, T. The optical fiber can be either single mode or multimode and the transducer can be either intrinsic or extrinsic. The input state, $\{I_0\}$, is transformed to an output state, $\{I\}$, according to Equation (1).

$${I} = T (X, {I_0}, {E_i})\qquad(1)$$

The modulated light signal is eventually detected by an optical detector, D, the output of which is processed with an inverse of the transfer function, T^{-1}, to determine the value of the desired parameter, X. Unfortunately, {I} is not just a function of X, but is also a function the environment around the transducer as measured by the environmental parameters, {E_i}. Changes in the input optical state, {I_0}, will also modulate the output. Any change in the inferred parameter, dX, is of course equivalent to $d[T^{-1}({I})]$. The total derivitive of T^{-1} is given by

$$d[T^{-1}({I})] = \frac{\partial T^{-1}}{\partial {I}}\, d{I} + \frac{\partial T^{-1}}{\partial {I_0}}\, d{I_0} + \frac{\partial T^{-1}}{\partial {E_i}}\, d{E_i}\qquad(2)$$

A well designed transducer should be used with an optical source, fiber and detector that minimize the second and third terms. At the very least, some technique should be provided to compensate for the errors that they introduce into the final estimation of the desired parameter, X. Optimization of fiber optic sensor design requires minimization of the final two terms in Equation (2), and many innovative concepts have been developed to perform that task.

2.2 Grating optics

An optical grating is a periodic structure that affects the state of an optical signal. The most well known type of grating is the diffraction grating whose action on a collimated beam of light of wavelength, λ, is shown in Figure 2.

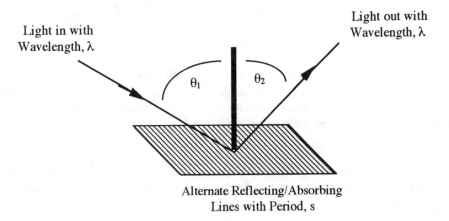

Light in with
Wavelength, λ

Light out with
Wavelength, λ

θ_1 θ_2

Alternate Reflecting/Absorbing
Lines with Period, s

Figure 2

The input and output angles are related via the grating equation, i.e.

$$s (\sin \theta_1 + \sin \theta_2) = m\, \lambda\qquad(3)$$

In this equation, angles are measured in the plane containing the incident angle and the normal to the grating surface. The order of the diffraction is defined by m which can be any integer, positive or negative. As can be seen by examining this equation, an optical grating can diffract an incoming light signal based upon its wavelength and the period of the line spacing on the grating. Depending on the optical configuration, a grating could then be used as part of a transducer whose transfer function served as a narrow band wavelength filter of an incoming optical signal and/or a propagation direction modulator.

Gratings can also be configured so as to modulate the intensity of optical signals. In order to do this, a pair of equivalent gratings must be used. An example of this type of modulation is shown in Figure 3.

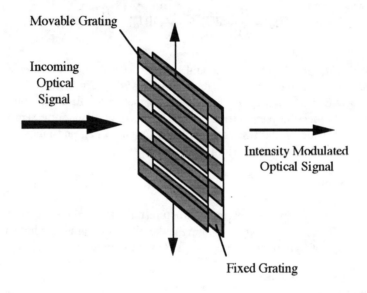

Figure 3

Finally, gratings can be formed by any periodic variation in an optical medium. An example of such a "phase grating" is shown in Figure 4. Other grating structures based on periodic index of refraction variations have been fabricated within optical fibers themselves [10] and offer great promise for the future of distributed strain sensing systems.

Transducers based upon gratings can therefore produce a number of different types of transfer function including intensity, I, wavelength, λ, and direction of propagation, $\mathbf{n}$.

2.3 Polarization optics

In classical terms, polarization is the direction of the electric field, $\mathbf{E}$, of an electromagnetic wave at a particular location in space. Although a rigorous discussion of polarization in quantum mechanical terms is beyond the scope of this paper, it should be pointed out that quantum effects are beginning to be explored for the purposes of optical communication (squeezed states) and it is likely that non-classical fiber optic

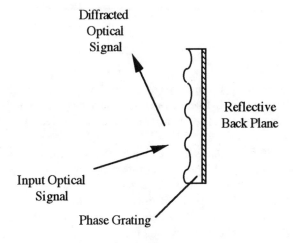

Diffracted
Optical
Signal

Reflective
Back Plane

Input Optical
Signal

Phase Grating

Figure 4

sensing techniques will also be developed. In this paper, however, classical electromagnetic theory and terminology will be generally used.

A snapshot in time of a linearly polarized electromagnetic wave is shown in Figure 5. As can be seen, associated with the wave is an electric field, **E**, and a magnetic

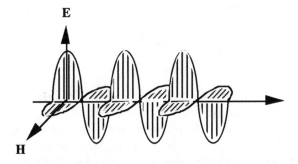

Travelling Electromagnetic Wave

Figure 5

field, **H**. The wave propagates in the direction of the Poynting vector, $\mathbf{S} = \mathbf{E} \times \mathbf{H}$. For visible light, the frequency of oscillation of the field is such ($\sim 10^{13}$ Hz) that it is not generally measured directly. Instead, the intensity of the wave, proportional to $\mathbf{E \cdot E}$, is typically measured.

A number of different formalisms exist for the analysis of polarized light. One of the more commonly used is Jones Calculus [2]. In this formalism, the state of polarization (SOP) of light is defined mathematically as a two dimensional vector, **a**, i.e.

$$\mathbf{a} = \begin{bmatrix} A_x\, e^{i\,\delta_x} \\ A_y\, e^{i\,\delta_y} \end{bmatrix}$$

(4)

where the light is assumed to be propagating in the **z** direction, having components proportional to the electrical field components in the **x** and **y** directions with amplitudes A_x, A_y and phases δ_x and δ_y. The intensity of the light is given by $\mathbf{a} \cdot \mathbf{a}$.

When the SOP of light is defined via its Jones vector, **a**, its modulation due to the passage of the light through optical elements such as waveplates, polarizers, etc., can be determined by multiplying the incident a vector by an appropriate two dimensional matrix representing that element. The forms of these matrices have been determined empirically and are well known. Of particular interest are the matrices representing the linear polarizer, $\mathbb{P}$, the linear retarder, $\mathbb{R}$, and the rotator, $\mathbb{T}$.

$$\mathbb{P} = \begin{bmatrix} \cos^2\theta & \sin\theta\cos\theta \\ \sin\theta\cos\theta & \sin^2\theta \end{bmatrix}$$

(5)

$$\mathbb{R} = \begin{bmatrix} e^{i\,\delta}\cos^2\theta + \sin^2\theta & (e^{i\,\delta}-1)\sin\theta\cos\theta \\ (e^{i\,\delta}-1)\sin\theta\cos\theta & e^{i\,\delta}\sin^2\theta + \cos^2\theta \end{bmatrix}$$

(6)

$$\mathbb{T} = \begin{bmatrix} \cos\theta & \sin\theta \\ -\sin\theta & \cos\theta \end{bmatrix}$$

(7)

In Equation (5), the angle θ represents the angle that the transmission axis of the polarizer makes with the **x** axis in the x-y plane. In Equation (6), the angle θ represents the angle between the fast axis of the linear retarder and the **x** axis while δ represents the relative phase retardation experienced between light polarized parallel to the fast axis of the retarder and light polarized parallel to its slow axis after they have both passed through it. Finally, the angle θ in Equation (7) represents a rotation around the **z** axis or propagation direction of the light.

The great benefit to utilizing a formalism such as Jones Calculus lies in the fact that input and output SOP's can be determined for passage of light through an arbitrary number of optical elements via simple matrix multiplication techniques, i.e.

$$\mathbf{a}' = \mathbb{M}_n \cdot \mathbb{M}_{n-1} \cdots \mathbb{M}_1 \cdot \mathbf{a}$$

(8)

When modulation of the SOP is used by the transduction mechanism, the transfer function associated with the modulation could involve the polarization direction, the phase, δ, and/or the overall intensity of the optical signal, depending upon the transducer details.

3. GRATING SENSORS

A number of different types of fiber optic sensors have been demonstrated whose transduction mechanisms involved the use of optical gratings of one sort or another. In general, these sensors have modulated either the wavelength or the intensity of the optical signal that passes through them. Both multimode and single mode sensors have shown considerable potential for continued development.

3.1 Multimode sensors

Fiber optic sensors utilizing optical gratings and multimode fiber technology were among the first to be developed. Over a period of time, the sophistication of device design has increased to the point that practical high accuracy sensors have now been developed.

3.1.1 Sensors based on grating intensity modulation

One of the more basic types of optical transducer mechanism is the one shown in Figure 3. In this case, the transfer function for the transducer is based on the relative position of two absorption gratings in very close proximity. If the grating period is very large compared to the wavelength of light in the incoming optical signal with equal areas of absorbing and transmissive surface, then the transmission through the opposed grating structure will vary from 50% when the gratings are in perfect alignment to 0 when their relative displacement is equal to 0.5 grating period. The transfer function for a structure of this type would then transform in input light intensity, I_0, into an output intensity, I, according to

$$I = T(x, I_0, s) = (1 - \frac{2x}{s}) I_0 \qquad 0 \le x \le \frac{s}{2} \tag{9}$$

where s is the grating period and x is the relative grating displacement. Any phenomenon, P, that can be used to affect the relative displacement of the opposed gratings $(x = x(P) = T^{-1}(I, I_0, s))$ can then be detected by monitoring I, i.e.

$$P = x^{-1} \{ \frac{s}{2} (1 - \frac{I}{I_0}) \} \tag{10}$$

The expression x^{-1} shown in Equation (10) is the inverse of the functional dependence of x on P.

An examination of Equation (10) reveals two direct sources of potential error and one indirect one. If the intensity of the optical signal input to the transducer varies, it will produce a change in the detected power, I. Application of Equation (10) to determine P will then result in an error in its inferred value since I_0 is assumed constant in that equation. In the same way, any variation in the grating period, s, will result in a similar error. Any change in the functional dependence of x upon P will create additional error. Finally, if the grating period is made small (comparable to the wavelength of light) to improve sensitivity (at the expense of dynamic range), wavelength effects due to diffraction could also introduce error. Development of practical sensors depends upon overcoming these problems.

One of the first fiber optic hydrophones developed was based upon the relative motion of two opposed gratings [5]. A schematic diagram of this type of sensor is shown in Figure 6. Light from a light emitting diode with peak emission at 820 nm

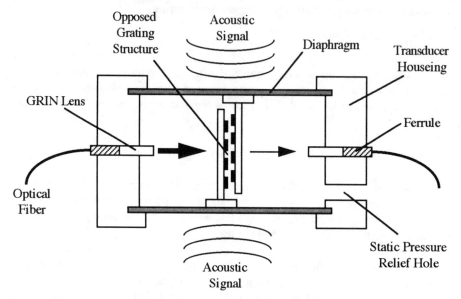

Opposed Grating Fiber Optic Hydrophone

Figure 6

was coupled into a 200 μm core silica optical fiber. In the transducer, the light was removed from the fiber, collimated by a GRIN lens and passed through the opposed grating structure which modulated the transmitted intensity based upon the acoustic signal impinging upon the transducer diaphragms. The modulated optical signal was then collected by a second GRIN lens and coupled into an output fiber and transmitted to a photodiode detector. The opposed grating structure consisted of 5 μm grating stripes deposited on microscope cover slips. The opposed gratings were aligned under a microscope and sealed at the edges with a soft RTV epoxy. An index matching oil was contained in the volume between the two gratings to reduce optical losses, minimize friction and to prevent the structure from collapsing under high static pressures. Plexiglass end caps were bonded to the structure which was then aligned within the transducer housing and bonded to the two rubber diaphragms.

The opposed grating hydrophone demonstrated its ability to detect very small acoustic signals both in the laboratory using and NRL G19 hyrophone calibrator with a Gould CH-17UT reference hydrophone and at the Sperry Corporation underwater test facility in Great Neck Long Island where it was submerged to a depth of 2.7 m and subjected to standard hydrophone tests under approximately free field conditions. A comparison of the detectecd signals from the opposed grating hydrophone and a standard piezoelectric hyrophone is shown in Figure 7. Both hydrophones detected gated

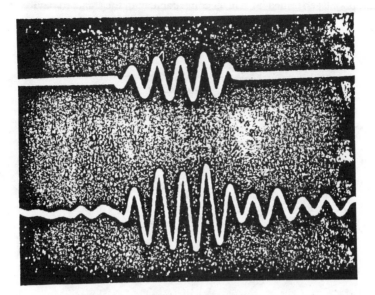

(a)

Acoustic SourceDrive Voltage (Top) vs Opposed
Grating Hydrophone Output (Bottom)

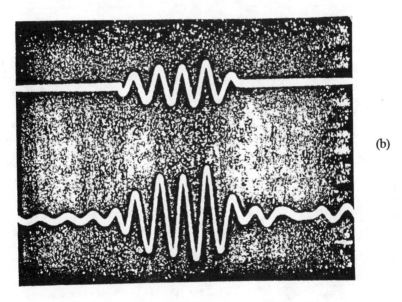

(b)

Acoustic Source Drive Voltage (Top) vs Standard
USRD Type F37 Hydrophone Output (Bottom)

Figure 7

acoustic signals at 1500 Hz with a gate width of 2.5 ms and repetition rate of 30
pulses/s. Figure 7 (a) shows the acoustic source drive voltage above the detected signal
from the optical hydrophone while Figure 7 (b) shows the acoustic source voltage above
the detected signal from a standard USRD Type F37 hydrophone. As can be seen, the

optical hydrophone performed at a level comparable with the standard device. Figure 8

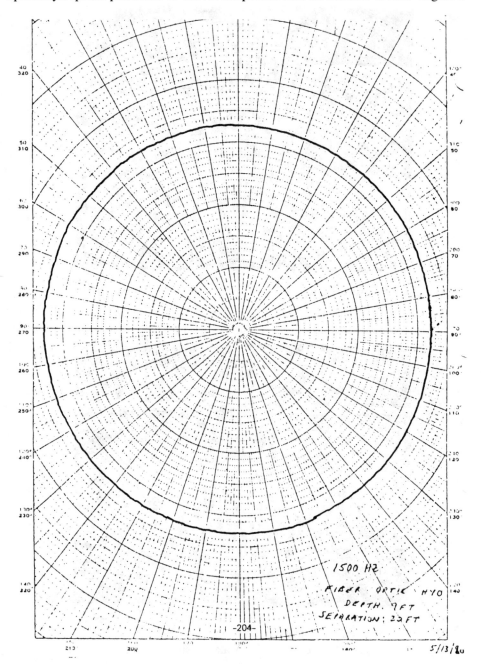

Directivity Response of Opposed Grating Hydrophone

Figure 8

shows directivity tests of the opposed grating hydrophone and as can be seen, the optical device displays omnidirectional behavior. Finally, tests of acoustic sensitivity demonstrated that the optical hydrophone could detect acoustic signal levels at or below sea state zero from 100 Hz to 1.2 kHz.

Although this early demonstration of a fiber optic sensor was very successful, several problems existed with the device. In particular, since no form of self referencing was used, intensity noise could enter the system through variable connector losses, transmission losses in the optical fiber, changes in optical source output or changes in photodetector sensitivity. In spite of that, this device did provide some of the first evidence that a fiber optic sensor could provide performance equivalent to an existing electrical sensor.

The opposed grating transduction mechanism has also been used to sense pressure [6]. The pressure sensor had an advantage over the earlier optical hydrophone in that the transducer utilized a self referencing technique. This is shown in Figure 9.

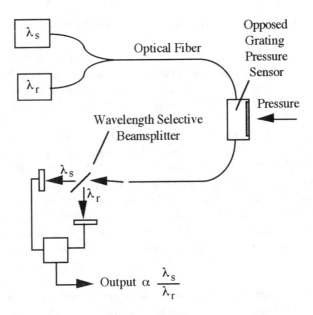

Self Referenced Opposed Grating Pressure Sensor

Figure 9

In this sensor, the opposed grating structure was fabricated so that the absorbing stripes would absorb at a signal wavelength, λ_s, but would be transmissive at some second reference wavelength, λ_r. If these two wavelengths are relatively close together, each will suffer equivalent non-transducer losses due to the optical fiber, connectors, etc. Dividing the detected intensity at the signal wavelength by the detected intensity at the reference wavelength will then in principle provide a measure of the transducer modulation and hence the pressure, independent of non-transducer intensity losses. A number of different sensing configurations can be used with this type of transducer. In Figure 9, light from two optical sources is continuously transmitted down the fiber, acted on by the transducer, and then separated into signal and reference components by a wavelength selective beamsplitter. The two signals are then processed electronically to provide a continuous output that is proportional to their ratio. An alternate technique having simpler optics but more complicated electronics involves alternately turning on each of the optical sources. In this case, only a single detector is required and the wavelength dependent beamsplitter is no longer needed. Control circuitry is required to co-ordinate the time sequencing of the optical sources and the processing of the detected signals.

This opposed grating transducer offered significant advantages over the hydrophone device in that it was inherently insensitive to intensity noise induced by environmental effects on the optical transmission path. It was, unfortunately, still subject to errors induced by changes in output of the optical sources over time, changes in optical source coupling into the optical fiber and changes in detector responsivity. In spite of its limitations, this sensor represents a practical method of fiber optic sensing for moderate accuracy applications.

3.12 Sensors based on grating wavelength modulation

Although a number of fiber optic sensors have been demonstrated that utilize intensity modulation, the problems of environmentally induced error in the signal remain extremely difficult to overcome. This limits sensors based on intensity to moderate sensitivity applications unless the transducer utilizes the intensity only as a carrier for information that is encoded via modulation in frequency. This is of course, the well known benefit of frequency modulation (FM) over amplitude modulation (AM) for radio broadcasting. Fiber optic sensors have been used to detect relatively low frequencies as evidenced by the optical hydrophone described in Section 3.1.1. If only frequency information is needed, then sensors of this type are adequate. Since amplitude information is also crucial for acoustic measurements, techniques other than straightforward intensity detection must be utilized to encode the signal. One of the more attractive methods from an optical standpoint is to modulate the frequency (wavelength) envelope of the optical carrier itself. If this is done, the problems associated with intensity sensing can be greatly reduced if not eliminated.

One of the first demonstrations of wavelength encoding for a fiber optic sensor showed that angular displacement could be encoded via wavelength modulation of a broadband optical signal [7]. In this case, as shown in Figure 10, a diffraction grating was attached to a shaft coupled to the object undergoing angular displacement.

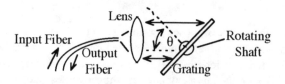

Angular Displacement Sensing with Wavelength Encoding

Figure 10

The operation of this transducer can be understood by examining Equation (3). In this equation, the sign convention for angular measurement is that angles on the same side of the normal as the incident angle are positive while those on the other side are negative. If the angle of incidence and the angle of diffraction are assumed to be the same (i.e. $\theta_1 = \theta_2 = \theta$), then the angle θ in Figure 10 can be identified as being equivalent to this angle and solved for in terms of the first diffracted order (m = 1)

$$\theta = \sin^{-1}(\frac{\lambda}{2\,s}) \tag{11}$$

This means that if a relatively broadband signal from a light emitting diode is coupled into the input fiber, it is then collimated by the lens and strikes the grating an an incidence angle, θ, that is equal to the angular displacement that one wishes to sense. The only light diffracted back along the incident path will be of wavelength, λ, according to Equation 11. This light will then be focused into the output fiber and transmitted to a signal processing location where its wavelength will be determined. Knowledge of λ and s then uniquely determine the required value of θ. The diffraction grating has functioned as a narrowband wavelength filter whose center wavelength is related to the angle of rotation. The transfer function, T, is given by

$$\lambda = T\,(\theta,\,s) = 2\,s\,\sin\theta \tag{12}$$

with the inverse transfer function given by

$$\theta = T^{-1}\,(\lambda, s) = \sin^{-1}\,(\frac{\lambda}{2\,s}) \tag{13}$$

As long as there is enough optical signal to be able to determine, λ, intensity noise should not affect the accuracy of this measurement. Wavelength noise, however, can enter the system through changes in the grating spacing with temperature, but this problem can be minimized through careful mechanical design.

Although it demonstrates many of the benefits of wavelength encoding, this fiber optic sensor is not in wide use and likely will not be due to the limited range of angles that it can sense. It did, however, lay the groundwork for the development of other grating based wavelength encoding fiber optic sensors.

One class of grating based fiber optic sensors that has demonstrated very high accuracy and insensitivity to changes in environmental parameters such as temperature and vibration [8,9] utilizes gratings whose period varies as a function of position. The basic transduction mechanism is shown in Figure 11.

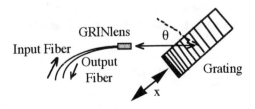

Figure 11

The grating period varies over the length of the grating according to

$$s = s_0 + s_1 x \tag{14}$$

Via a mechanical linkage, the grating is mechanically attached to the object whose linear displacement is to be determined and allowed to move with the object. The other optical elements, however, remain fixed. This implies that the parameter, x, in Equation (14) can be identified with the linear displacement to be determined. The transfer function of a sensor of this type is then given by

$$\lambda = T(\theta,x) = (2 s_0 \sin \theta) + (2 s_1 \sin \theta) x \tag{15}$$

with the inverse transfer function given by

$$x = T^{-1}(\lambda,\theta) = (-\frac{s_0}{s_1}) + (\frac{1}{2 s_1 \sin \theta}) \lambda \tag{16}$$

As can be seen, the linear displacement, x, can be uniquely determined from a measurement of λ if the variation of grating period with displacement and diffraction angle are known. Errors can occur in the inferred value of x if the value of θ varies, e.g. under high vibration levels, if the grating period varies, e.g. with temperature, or if the measured value of λ is in error, e.g. due to the signal processing details of determining the center wavelength with extremely high accuracy of a band pass filter of a non-uniform and time/temperature dependent optical source intensity (λ) envelope.

To overcome potential performance problems, advanced versions the device reported in the literature [9] have utilized careful mechanical design techniques to minimize the sensitivity of θ to vibration and used a grating substrate with a very low coefficient of thermal expansion to minimize the sensitivity of the grating period, s(x), to changes in temperature. Errors in determining the diffracted wavelength were minimized through the use of advanced grating fabrication technology and opto-mechanical design to produce an effective wavelength bandpass of ~4 nm (full width half maximum). A compact, accurate and reduced cost method of wavelength measurement was also created during the sensor development. The technique utilized a low cost dual photodiode that had two outputs corresponding to overlapping but different wavelength resposes. This is shown in Figure 12. As can be seen, the two responsivities have roughly the same shape, but different center wavelengths and different maximum values. Although developed for other applications, the benefits of using the dual photodiode device for processing the output of the variable period grating sensor were considerable [23]. The functioning of the device may be understood by modeling the responsivities as simple exponential decays in the measurement region. It will also be assumed that each curve has the same decay factor, K. The detected signal currents from each photodiode would then be

$$i_1 = R_1 e^{K(\lambda_1 - \lambda)} I$$

$$i_2 = R_2 e^{K(\lambda - \lambda_2)} I \tag{17}$$

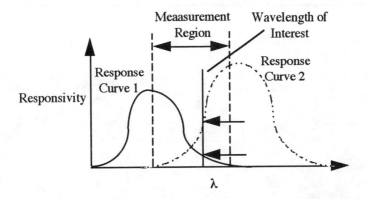

Overlapping Responsivity Curves from Dual Photodiode

Figure 12

where the I is the optical intensity and (R_1, λ_1) and (R_2, λ_2) are the respective photodiode constants. In the measurement region, $\lambda_1 < \lambda < \lambda_2$. Taking the ratio of the two detected currents and rearranging terms, the wavelength, λ, can be determined, i.e.

$$\lambda = -\frac{1}{2K} \ln \left[\frac{R_2}{R_1} e^{K(\lambda_2 - \lambda_1)} \right] - \frac{1}{2K} \ln \left[\frac{i_1}{i_2} \right] \tag{18}$$

The wavelength is then uniquely determined from a measurement of the two photocurrents, i_1 and i_2. Unfortunately, the dual photodiode configuration is quite temperature sensitive, so that in this model, K, λ_1, λ_2, R_1, and R_2 would all be temperature dependent. In order to compensate for this temperature dependence, as part of the sensor operation, a real time calibration procedure was periodically carried out in which the optical source would be turned off and the V - I curve of one of the photodiodes would be measured in the absence of illumination. Since the characteristics of this curve are very temperature dependent, this procedure allowed the temperature of the dual photodiode to be determined and its responsivities to be calculated based on prior device characterization data. Very high accuracy wavelength measurement accuracies have been obtained with this technique [24].

A prototype version of the device (without the dual photdiode wavelength detection) was extensively tested by a large aircraft manufacturer and was found to have considerable promise for application in the flight control system in future fly-by-light aircraft. This device is shown in Figure 13. The device operated over a linear displacement range of 5 cm, with overtravel of ±1 cm, resolution of 0.1% full scale and accuracy of 0.25% over the temperature range -55 C to +85 C.

3.2 Single mode grating sensors

One of great potential and as yet unrealized benefits of optical fiber sensors is their potential ability to provide distributed measurements of one or more parameters over very long gauge lengths with moderate to high accuracy. Although a number of attempts have been made to develop fiber optic sensors to perform this task, none has

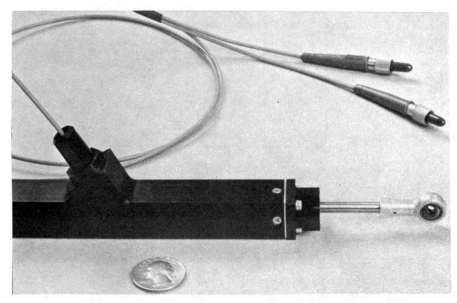

Prototype Variable Period Grating Linear Displacement Sensor

Figure 13

seemed to offer the potential for widespread and general use. The recent development of the ability to create in-line gratings in optical fibers based upon controlled index of refraction variations [10] offers the best opportunity to date to fulfill the promise of a general distributed sensing technique.

The creation of in-fiber Bragg gratings depends upon the well known fact that a periodic intensity is produced by two coherent light beams when they intersect each other at an angle [25]. This situation is shown in Figure 14. If the intersection angle of the two beams is 2θ, then the period of the interference maxima produced within the intersection region on a plane perpendicular to the axis bisecting the angle between the two beams is given by

$$s = \frac{\lambda}{2 \sin \theta} \tag{19}$$

This technique has been used in a configuration in which the substrate is a germanium doped single mode optical fiber exposed to the interference of two intersecting beams of very high intensity coherent ultraviolet light. The interaction of the interference pattern with the germanium doped core fiber produces a periodic index of refraction variation longitudinally along the fiber length. These longitudinal index gratings have the property of acting as a reflector for modes in the fiber with a wavelength that is twice the period [26], s, i.e.

$$\lambda_b = \frac{\lambda_i \, n_e}{\sin \theta} \tag{20}$$

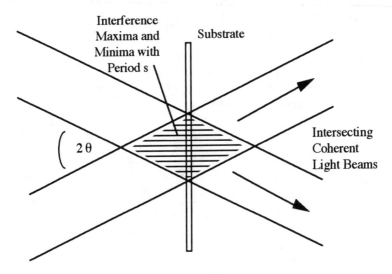

Generation of Periodic Intensity Distribution Through
Interference of Two Coherent Light Beams

Figure 14

where λ_b is the wavelength that will be backreflected by the longitudinal grating, λ_i is the wavelength of the UV light that created the grating and θ is one half the intersection angle between the two beams.

An example of a fiber sensing system using Bragg gratings is shown in Figure 15. In this case, light from a broadband source is coupled into the sensing fiber which sees strains and index of refraction changes at four grating locations. The light leaving the fiber is then analyzed as to its spectral intensity distribution as shown in Figure 16. As can be seen, there are a number of notches in the spectrum, each corresponding to one of the grating regions. If the temperature is constant, then the wavelength shift of a given notch will correspond to the longitudinal strain seen by its respective grating according to [26]

$$\frac{\Delta\lambda_b}{\lambda_b} = (1 - p_e)\, \varepsilon \tag{21}$$

where the quantity p_e is an effective photoelastic constant given by

$$p_e = (\frac{n^2}{2})\, (p_{12} - v\, (p_{11} + p_{12})) \tag{22}$$

where n is the index of refraction of the fiber core, p_{12} and p_{11} are components of the strain optic tensor and v is Poisson's ratio. A knowledge of the wavelength of the notches and monitoring their shifts as a function of time can allow determination of the changing strain levels seen at a particular location along the fiber. For strain sensing (at constant temperature) the transfer function for a Bragg grating sensor is given by

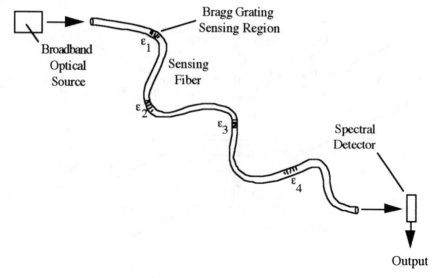

Distributed Strain Sensing with Bragg Gratings

Figure 15

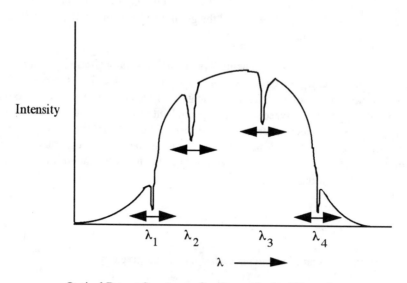

Optical Power Spectrum after Transmission Through
a Bragg Sensing Fiber with Four Sensitive Regions

Figure 16

$$\Delta\lambda_b = T\left(\varepsilon, \lambda_b, p_e\right) = \lambda_b \left(1 - p_e\right) \varepsilon \qquad (23)$$

while the inverse transfer function is given by

$$\varepsilon = T^{-1}(\Delta\lambda_b, \lambda_b, p_e) = \frac{\Delta\lambda_b}{\lambda_b(1-p_e)} \tag{24}$$

If there is no stress induced strain, equivalent expressions can be developed which allow the fiber to monitor temperature distributions.

The in-fiber Bragg grating sensing technique is one of the more promising fiber optic sensor development areas. It is an intrinsic sensor (light never leaves the fiber), it can be utilized in a transmissive mode, and a large number of sensing locations can be produced on a single length of fiber. There are, however, a number of factors that are holding back widespread use of the technology. Either a compact broadband optical source or a widely tunable semiconductor laser is needed to excite the sensing fiber. Such a source does not exist at present. An accurate and cost effective method of carrying out spectral analysis is also needed. Finally, an effective signal conditioning technique is required for separation of variables so that thermal and stress induced strain effects can be distinguished in a straightforward manner.

4. POLARIMETRIC SENSORS

A number of different fiber optic sensors have been developed that rely on modulation of the polarization state of an optical signal to encode information about a parameter of interest. In particular, the polarization state can be modified by passage through a polarizer, $\mathbb{P}$, a linear retarder, $\mathbb{R}$, and a rotator, $\mathbb{T}$. Analysis of the operation of these kinds of sensors can then be carried out using Jones Calculus and Equations (4-7), remembering that the the detected optical power is obtained from the final Jones vector according to

$$I = a_x^2 + a_y^2 \quad \text{where} \quad \mathbf{a} = \begin{bmatrix} a_x \\ a_y \end{bmatrix} \tag{25}$$

4.1 Multimode sensors

One of the simplest sensors that utilizes polarization modulation measures angular displacement. This transducer consists of two polarizers, one of which is fixed and second one that is attached to the axis of rotation around which the angular displacement is to be measured. This is shown in Figure 17. If unpolarized light of intensity I_0 is passed through the transducer, the emerging polarization state will be given by

$$\mathbf{a} = \mathbb{P}(\theta)\,\mathbb{P}(0)\left[\frac{I_0}{2}\right]^{\frac{1}{2}}\begin{bmatrix} 1 \\ 1 \end{bmatrix} \tag{26}$$

where the argument of the polarizer matrix is the angle between the transmission axis and some reference axis. Carrying out this calculation provides the transfer function between the input and output optical power, i.e.

$$I = T(\theta, I_0) = \frac{I_0}{2}\cos^2\theta \tag{27}$$

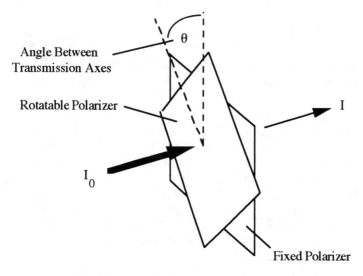

Intensity Modulation via Angular Displacement
of the Transmission Axes of Two Polarizers

Figure 17

The inverse of the transfer function is then given by

$$\theta = T^{-1} (I, I_0) = \cos^{-1} \left(\left[\frac{2\,I}{I_0} \right]^{\frac{1}{2}} \right)$$

(28)

The major attractiveness of this approach is its simplicity, both in implementation and analysis. Laboratory demonstrations of sensors using this technology have shown that moderate accuracies can be obtained when it is used to measure angular displacements [12]. The devices are, however, still sensitive to errors created by intensity fluctuations from the optical source and environmental conditions.

A method of polarization modulation which has been more intensively investigated involves the use of the photoelastic effect to modulate the polarization state. A typical optical configuration for a transducer of this type is shown in Figure 18. An unpolarized optical signal is first transmitted through an input polarizer with transmission axis at $\pi/4$ to the axis of the applied stress, σ. Next it passes through a linear retarder (retardation $\delta = \pi/2$) with fast axis parallel to the axis of applied stress. It then is modulated by a stress induced retardation, δ_σ, and passed through an output polarizer with transmission axis at $\pi/4$ to the stress axis. The stress induced retardation is given by [2]

$$\delta_\sigma = \frac{2\,\pi\,t\,C_\sigma}{\lambda}\,\sigma$$

(29)

where t is the thickness of the photoelastic element, C_σ is the photoelastic constant for the material and λ is the wavelength. The retardation is then directly proportional to the applied stress. The relationship between the input and output optical powers can also be

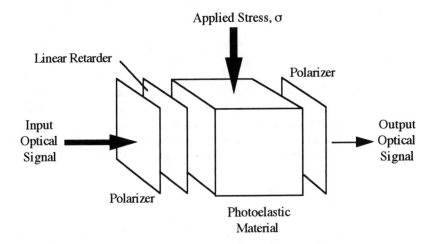

Light Signal Transmitted Through an Optical
Transducer Based on the Photoelastic Effect

Figure 18

analyzed for this configuration of optical elements using Jones Calculus. The
transmitted Jones vector is calculated from

$$\mathbf{a} = \begin{bmatrix} a_x \\ a_y \end{bmatrix} = \left[\frac{I_0}{2}\right]^{\frac{1}{2}} \mathbb{P}\left(\frac{\pi}{4}\right) \mathbb{R}\left(0, \delta_\sigma\right) \mathbb{R}\left(0, \frac{\pi}{2}\right) \mathbb{P}\left(\frac{\pi}{4}\right) \begin{bmatrix} 1 \\ 1 \end{bmatrix}$$

(30)

where the argument of the polarization matrices is the angle that their transmission axis
makes with the axis of the applied stress and the arguments of the retarder matrices are
the angles that their fast axes make with the axis of the applied stress and the amount of
retardation that they produce. Carrying out the calculation yields the transfer function for
this transducer, i.e.

$$I = T\left(\sigma, I_0, t, C_\sigma, \lambda\right) = \frac{I_0}{4}\left[1 + \sin\left(\frac{2\pi t C_\sigma}{\lambda}\sigma\right)\right]$$

(31)

with the inverse of the transfer function taking the form

$$\sigma = T^{-1}\left(I, I_0, t, C_\sigma, \lambda\right) = \frac{\lambda}{2\pi t C_\sigma}\sin^{-1}\left(\frac{4I}{I_0} - 1\right)$$

(32)

Equations (31,32) simplify greatly in the limit of small σ, and reduce to

$$I = T\left(\sigma, I_0, t, C_\sigma, \lambda\right) = \left(\frac{2\pi t C_\sigma I_0}{4\lambda}\right)\sigma + \left(\frac{I_0}{4}\right)$$

(33)

and

$$\sigma = T^{-1}(I, I_0, t, C_\sigma, \lambda) = (\frac{4\lambda}{2\pi t C_\sigma I_0}) I - (\frac{\lambda}{2\pi t C_\sigma}) \tag{34}$$

In this limit, the relationship between applied stress and optical signal is seen to be linear.

One of the first fiber optic sensors using a photoelastic element to modulate the polarization state of light passing through it was a hydrophone [27]. The device was fabricated using the same housing as shown schematically in Figure 6. The opposed grating transducer element was replaced by a photoelastic element of polyerethane rubber with polarizers and a bias waveplate bonded directly to it as shown in Figure 19.

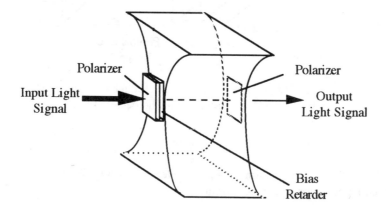

Transducer Element for Photoelastic Hydrophone

Figure 19

The hydrophone provided compensation for optical source intensity drift by having two output signals, one corresponding to an output polarizer with transmission axis at $+\pi/4$ and one corresponding to an output polarizer having transmission axis at $-\pi/4$. This was done by passing the output from the hydrophone transducer element (minus the output polarizer) through a half wave plate to rotate the $\pm\pi/4$ light polarization directions to $\pi/2$ and 0 [27]. The light was then passed through a polarizing beamsplitter that separated out the two components. The two output signals were then coupled into separate output fibers and finally converted into electrical signals via photodiode detectors. The two optical signals had intensities given by

$$I_\pm = \frac{I_0}{4}\left[1 \pm \sin\left(\frac{2\pi t C_\sigma}{\lambda}\sigma\right)\right] \tag{35}$$

The optical detection scheme functioned so as to provide a voltage output, V_s, that was proportional (with proportionality constant k_s) to the difference over sum of the two optical signals, i.e.

$$V_s = k_s \sin\left(\frac{2\pi t C_\sigma}{\lambda}\sigma\right) \tag{36}$$

where as can be seen, the input optical intensity is not present. The signal to noise ratio improvement that results from using this detection scheme is shown in Figure 20.

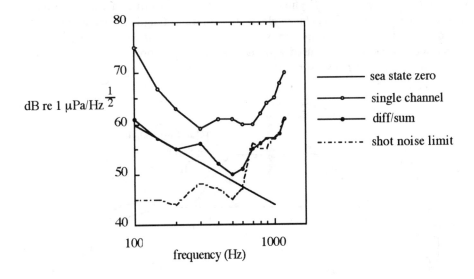

Photoelastic Hydrophone Sensitivities

Figure 20

Although a laser source was used in these sensitivity measurements, the results clearly show an improvement of ~10 dB in signal to noise ratio for the difference over sum technique as opposed to simple intensity detection from a single channel. Furthermore, calculations assuming a shot noise limit indicated that the device would have a sensitivity well below sea state zero if the laser were replaced a shot noise limited source.

The success of this hydrophone led to the development of a four hydrophone array based on the technology that was developed under an NRL contract [28]. The final report for that program provides a detailed discussion of the steps required to design and fabricate a fiber optic sensor suitable for field testing.

Other sensors have been demonstrated with photoelastic transduction. A simple accelerometer was shown to be able to detect ac accelerations as small as ~1 µg [16], but work on it was discontinued due to the fact that it was sensitive to accelerations other than in the direction of interest. A more successful sensor was used to measure pressure [13]. It was similar to the photoelastic hydrophone [27] in terms of having dual outputs and difference/sum detection. Differences included a single Be-Cu diaphragm as opposed to two rubber diaphragms and a photoelastic element of soda lime glass instead of polyurethane rubber. The device was shown to have minimum hysteresis and excellent sensitivity for the pressure range 0 - 3.5 MPa. Advanced versions of this type of pressure sensor with a novel compensation scheme to provide high accuracy and repeatability have been developed in the People's Republic of China [14,15] where they are now being used to measure the quantity of oil in very large storage tanks.

4.2 Single mode sensors

Although a large number of multimode fiber optic sensors have been developed that are based on the photoelastic effect, it is also true that a single mode optical fiber can act as a photoelastic element of arbitrary length. Such sensors offer some of the same benefits in terms of geometric versatility as single mode interferometric sensors but can used with more simple signal processing techniques in certain applications. Such a sensor is shown schematically in Figure 21.

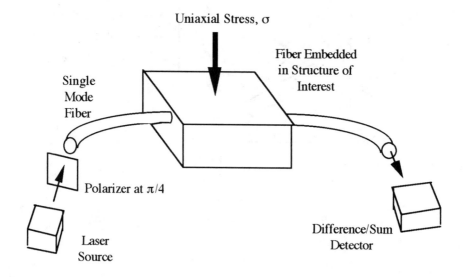

Single Mode Polarimetric Sensor for Internal Stress Monitoring

Figure 21

In this case a laser source is used to excite a single mode fiber being utilized as an embedded probe to monitor internal stress levels in a rigid structure. This is a particularly useful technique for composite structures where considerable use has been made of it [29].

One of the earliest sensors of this type was a hydrophone that demonstrated good sensitivity [17] but was not pursued further due to the excellent results being obtained with interferometric fiber optic hydrophones at that time. More recently, a sensor based on the technique has been used to monitor vibrations [18]. It is for low accuracy, "quick and dirty" kinds of stress/strain vibration measurement where sensors of this type are seeing the most use.

4.3 Faraday effect polarimetric sensors

Perhaps the most promising transduction technique using polarization modulation is based upon materials which exhibit the Faraday effect. A schematic diagram which illustrates the effect is shown in Figure 22. Linearly polarized light enters the Faraday material. Depending upon the distance traveled in the material and the component of applied magnetic field, **H**, parallel to the light propagation direction, the

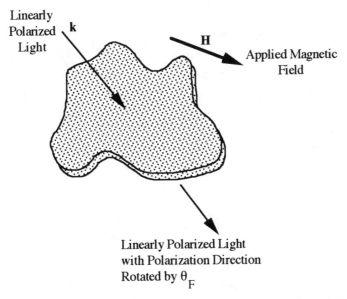

Faraday Material Modulating a Linear Polarization State Through Rotation

Figure 22

polarization of the light will be rotated through an angle, θ_F, given by

$$\theta_F = V \ (\mathbf{k} \cdot \mathbf{H}) \ t \tag{37}$$

where V is the Verdet constant of the material, $\mathbf{k}$ is a unit vector in the direction of light propagation, $\mathbf{H}$ is the applied magnetic field and t is the distance that the light travels through the material. As can be seen, the rotation is directly proportional to the applied field so that no rotation exists in the absence of the field. The effect of the applied field is to order the microscopic magnetic moments that exist within the material to create some net average moment. It is the non-zero average moment that exists with the field.

One of the materials that exhibits the Faraday effect is glass. For that reason, even though the Verdet constant for glass is small, the long gauge length that can be obtained through the use of a single mode fiber makes measurements of magnetic field possible. In this case the polarization rotation at the output of the fiber represents the path integral along the fiber of (V$\mathbf{H}$·d$\mathbf{s}$) [22]. There is considerable interest in applying this technique in large transformers by winding a sensing optical fiber in parallel with the electrical conductors. More work needs to be done, however, before such sensors have been developed enough to see wide use.

Another, more immediately practical, class of fiber optic sensors using Faraday effect transduction is based on small volumes of special materials that have large Verdet constants. One such material is yittrium iron garnet. This material has been used in a device to sense rotary speed, a tachometer [20,21]. The operation of the device is shown in Figure 23.

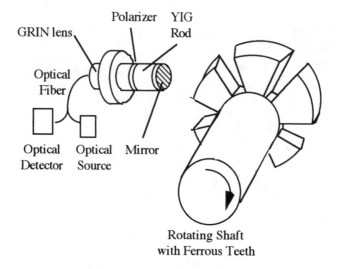

Multimode Fiber Optic Tachometer based on the Faraday Effect

Figure 23

In this sensor, light from a single optical fiber is collimated by a GRIN lens, passed through a polarizer, then a length of yitrium iron garnet rod, reflected by a mirror, passed back through the garnet rod and the polarizer and finally coupled back into the fiber by the lens. A permanent ring magnet is used to bias the garnet material so that it produces a $\pi/8$ value for θ_F. Since the light passes twice through the garnet and since the Faraday effect is non-reciprocal, the total rotation experienced by the light is $\pi/4$. If a ferrous target (such as a tooth attached to a rotating shaft) is brought into close proximity to the sensing head and the ring magnet, the field seen by the garnet rod along its light transmission axis will change by an amount ΔH. The detected optical power will then be given by

$$I = \frac{I_0}{2} \cos^2 \left(\frac{\pi}{4} + 2 V t \Delta H\right)$$
(38)

which in the limit of small ΔH reduces to

$$I = T(\Delta H, H_0, I_0, V, t) = (I_0 V t) \Delta H + \frac{I_0}{4}$$
(39)

where H_0 is the static bias field provided by the ring magnet. The inverse transfer function given by

$$\Delta H = T^{-1}(I, H_0, I_0, V, t) = \left(\frac{1}{I_0 V t}\right) I + \frac{1}{4 V t}$$
(40)

The frequency of the detected optical power will then be equivalent to the rotational speed of the shaft in question multiplied by the number of ferrous teeth.

The material yittrium iron garnet, while satisfactory in many respects in terms of performance, is too expensive to be used in fiber optic sensors. For that reason, more recent sensor activities have focused upon bismuth doped yitrium iron garnet films similar to those used in magneto-optic recording. These films exhibit large Faraday rotations and can have curie temperatures up to 325 C. Fiber optic proximity sensors have been built using these films [19].

The behavior of thin bismuth doped yitrium iron garnet films is significantly different from that exhibited in glass or yittrium iron garnet rods under applied magnetic fields. It is not evident from the technical literature that this difference is understood by most fiber optic sensor developers using these films.

A thin film of bismuth doped yittrium iron garnet consists of a large number of magnetic domains each with a magnetic moment normal to the surface of the film. A typical configuration for such a film is shown in Figure 24.

Domain Structure of Bismuth Doped Garnet Film (400 X)

Figure 24

Each adjacent domain has its magnetic moment in the opposite direction, i.e. light passing through one domain will have its polarization rotated by $+\theta_F$ which light passing through the adjacent domain will have its polarization rotated by $-\theta_F$. These magnetic moments exist with no field present so that polarized light passing through the film will always experience a rotation. The volumes of the two types of domains are equivalent, however, so that equal amounts of positive and negative rotation occur. Under the application of an applied field normal to the surface, the relative volumes of the two types of domain change but the magnetic moments within each volume do not change. Understanding this effect is critical in order to be able to design a fiber optic sensor using these films operating in reflection with a single polarizer.

The optical configuration of a thin film Faraday sensor is shown in Figure 25.

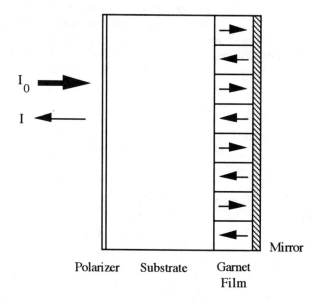

Bismuth Doped Garnet Film with No Applied Field

Figure 25

The volume of magnetic domains with magnetic moments facing to the right are defined as V_+ while the ones with left facing moments have volume V_-. The optical signals passing through the $V\pm$ volumes will experience rotations $\pm\theta_F$. The detected optical intensities will then then be

$$I = I_+ + I_- = \frac{I_0}{2} \frac{V_+}{V_+ + V_-} \cos^2(\theta_F) + \frac{I_0}{2} \frac{V_-}{V_+ + V_-} \cos^2(-\theta_F) = \frac{I_0}{2} \cos^2\theta_F \quad (41)$$

As can be seen, although the domain volumes are equal, their relative sizes do not affect the magnitude of the optical power returned from the transducer. If a magnetic field, magnitude H, is applied parallel to the V_+ magnetic moments, the V_+ volume increases while the V_- volume decreases as shown in Figure 26. Although the relative volumes have changed, the amount of rotation seen by light passing through the volumes remains the same as before, and Equation (41) remains an accurate equation. It is evident then that no modulation will result from fields applied normal to the film and the device will not function as transducer, i.e. its output will be constant.

Modulation will occur, however, if the positions of the polarizer and the mirror are reversed on the film as shown in Figure 27. If the volumes are assumed to change linearly with applied field (when one is not very much greater in magnitude than the the other) and if the total volume is given as V_T and some saturation magnetic field is given as H_s, then

$$V_\pm = \frac{V_T}{2} (1 \pm \frac{H}{H_s}) \quad (42)$$

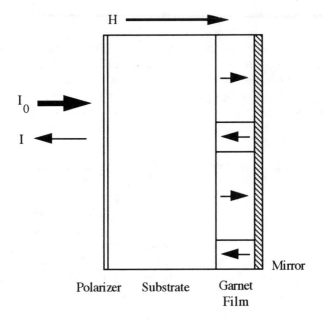

Bismuth Doped Garnet Film with Applied Field

Figure 26

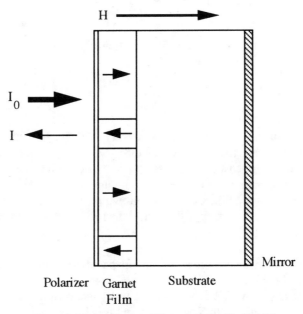

Bismuth Doped Garnet Film with Applied Field
and Mirror on Substrate Side

Figure 27

In this case, a fraction of the light input to the film passes through the plus volume and is rotated by an amount $+\theta_F$, it then travels to the substrate, is reflected, passes back through the substrate and returns through the Faraday film. A fraction of this reflected light passes through the plus volume again and receives a second rotation of $+\theta_F$. The remainder of the light passes through the minus volume and has its rotation cancelled. In a similar way, a fraction of the light initially passing through the minus volume ends up with a total rotation of $-2\theta_F$ while the remainder ends up with no rotation. The total optical signal transmitted through the transducer for the case shown in Figure 27 is then

$$I = \frac{I_0}{2} \frac{V_+^2}{V_T^2} \cos^2 (2\theta_F) + \frac{I_0}{2} \frac{2 V_+ V_-}{V_T^2} + \frac{I_0}{2} \frac{V_-^2}{V_T^2} \cos^2 (-2\theta_F) \tag{43}$$

where the first term represents the intensity from the light passing only through the plus volume, the middle term represents the intensity from light that passed through both plus and minus volumes to become unrotated and the last term represents light that only passed through the minus volume. If we combine this expression with the linear dependence on applied field of Equation (42), the dependence of the transmitted intensity on applied field becomes

$$I = T(H, I_0, \theta_F, H_s) = \frac{I_0}{4} [\cos^2(2\theta_F) + 1] + \frac{I_0}{4 H_s^2} [\cos^2(2\theta_F) - 1] H^2 \tag{44}$$

with the inverse transfer function having the form

$$H = T^{-1}(I, I_0, \theta_F, H_s) = \left[\left(\frac{4 H_s^2}{I_0 (\cos^2(2\theta_F) - 1)} \right) I - \left(\frac{H_s^2 (\cos^2(2\theta_F) + 1)}{\cos^2(2\theta_F) - 1} \right) \right]^{\frac{1}{2}} \tag{45}$$

If θ_F is chosen to be $\pi/4$ then Equation (45) simplifies to

$$H = T^{-1}(I, I_0, \theta_F, H_s) = \left[\left(\frac{4 H_s^2}{I_0} \right) I - H_s^2 \right]^{\frac{1}{2}} \tag{46}$$

The basic transfer function of Equation (44) is very different from the result for the general Faraday rotator case given by Equation (38), and it is important to understand the difference when designing sensors using these films. Sensors can be designed to operate quite satisfactorily in transmission if the transmission axes of the input and output polarizers are not parallel. Films which are grown on both sides of substrates will function in reflection, since mixing will occur in the two layers with their uncorrelated domain patterns.

In terms of practical application, sensors based upon the Faraday effect offer the most promise of the variety of polarization sensing techniques available. Successful testing of speed sensors and torque sensors using thin Faraday films has been carried out by major manufacturers of gas turbine engines and it is likely that sensors of this type will be used for aerospace applications within the next few years. A packaged thin film Faraday sensor (minus its bias magnet) is shown in Figure 28.

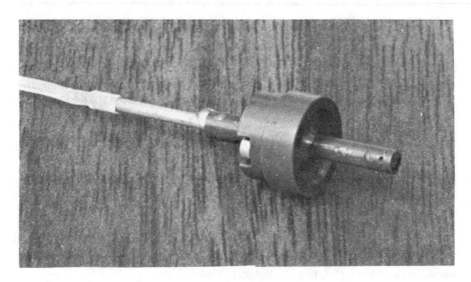

Packaged Thin Film Faraday Speed Sensor

Figure 28

5.   SUMMARY AND CONCLUSIONS

For the past several years, considerable activity has taken place aimed at producing optical fiber sensors for practical application. The results of those activities are beginning to be seen. Grating based sensors and polarimetric sensors have evolved from simple beginnings to the point at which sophisticated devices are now available. In particular, a linear position sensor using a grating with a variable period, a multimode photoelastic pressure sensor for oil tank quantity measurement and thin film Faraday sensors for speed and torque measurement have been developed that offer the potential of widespread use in the not so distant future.

6.   REFERENCES

1. *Proc. SPIE **412, 478, 566, 718, 838, 985, 1169, 1367, 1584**, Fiber Optic and Laser Sensors I-IX*, 1983-1991.
2. E. Udd, Editor, <u>*Fiber Optic Sensors: An Introduction for Engineers and Scientists,*</u> John Wiley & Sons, New York, 1991.
3. B. Culshaw, <u>*Optical Fibre Sensing and Signal Processing*</u>, Peter Peregrinus, London, 1984.
4. D.A. Krohn, <u>*Fiber Optic Sensors Fundamentals and Applications*</u>, Instrument Society of America, Research Triangle Park, North Carolina, 1988.
5. W.B. Spillman, Jr. and D.H. McMahon, "Schlieren Multimode Fiber-Optic Hydrophone", *Appl. Phys. Lett.* **37(2)**, 145-147, 1980.
6. B.E. Jones and R.A. Spoonser, "An Optical Fibre Pressure Sensor Using a Holographic Shutter Modulator with Two Wavelength Intensity Referencing", *Proc. 2nd International Conference on Optical Fibre Sensors*, Stuttgart, 1984.

7. C.R. Batchellor, J.P. Dakin and D.A.J. Pierce, "Fibre Optic Mechanical Sensors for Aerospace Applications", *Proc. SPIE 838, Fiber Optic and Laser Sensors V*, San Diego, 169-172, 1987.

8. W.B. Spillman, Jr., D.R. Patriquin and D.H. Crowne, "Fiber Optic Linear Displacement Sensor Based Upon a Variable Period Diffraction Grating", *Appl. Opt. 28(17)*, 3550-3552, 1989.

9. D.H. Crowne, D.R. Patriquin and W.B. Spillman, Jr., "Wavelength Encoded Fiber Optic Position Sensor", *Proc. SPIE 1169, Fiber Optic and Laser Sensors VIII*, Cambridge, 442-452, 1989.

10. G. Meltz, W.W. Morey and W.H. Glenn, "Formation of Bragg Gratings in Optical Fibers by a Transverse Holographic Method", *Opt. Lett. 14(15)*, 823-825,1989.

11. H.D. Simonsen, R. Paetsch and J.R. Dunphy, "Fiber Bragg Grating Sensor Demonstration in Glass-Fiber Reinforced Polyester Composite", *Proc. 1st European Conf. on Smart Structures and Materials*, Glasgow, 73-76, 1992.

12. J.T. Newmaster et al, "Remote Fiber Optic Sensors for Angular Orientation", *Proc. SPIE 838, Fiber Optic and Laser Sensors V*, San Diego, 28-38, 1987.

13. W.B. Spillman, Jr., "Multimode Fiber Optic Pressure Sensor Based on the Photoelastic Effect", *Opt. Lett. 7*, 388-391, 1982.

14. A.B. Wang, *High Accuracy Fiber Optic Sensing Oil Storage Measuring System*, Ph.D. Dissertation, Dalian Institute of Technology, 1989.

15. A.B. Wang, "Λ Compensation Method for an Optical Intensity Detector using Polarized Light", *People's Republic of China Invention and Patent Report 5(2), No. 147*, 1989.

16. W.B. Spillman, Jr., "Multimode Fiber-Optic Accelerometer Based on the Photoelastic Effect", *Appl. Opt. 21*, 2653-2655, 1982.

17. J.D. Beasley et al, "A Sensitive Polarization Hydrophone", *Proc. SPIE 412, Fiber Optic and Laser Sensors*, Arlington, 136-142, 1983.

18. W.B. Spillman, Jr. and B.R. Kline, "Fiber Optic Vibration Sensors for Structural Control Applications", *Proc. Damping '89 Conf. Vol. III*, West Palm Beach, ICA-1 - ICA-21, 1989.

19. L.B. Maurice et al, "Low Cost Binary Proximity Sensor for Automotive Applications", *Proc. SPIE 1173*, Fiber Optic Systems for Mobile Platforms III, Cambridge, 75-83, 1989.

20. B.J. Zook and C.R. Pollock, "Fiber Optic Tachometer based on the Faraday Effect", *Appl. Opt. 28(11)*, 1991-1994, 1989.

21. B.J. Zook, C.R. Pollock, and J.A. Morris, "Fiber Optic Transducer using Faraday Effect", U.S. Patent # 4,947,035, 1990.

22. A.J. Rogers, "Optical Measurement of Current and Voltage on Power Systems", *IEE Power Applications 2(4)*, 120-124, 1979.

23. M.C. Hutley, private communication.

24. D.H. Crowne, "Temperature-Compensated Signal Processor for Narrowband Wavelength Encoded Sensors", to be published in *Proc. SPIE 1795, Fiber Optics and Laser Sensors X*, Boston, 1992.

25. M.C. Hutley, *Diffraction Gratings*, Academic Press, New York, 1982.

26. W.W. Morey, G. Meltz and W.H. Glenn, "Fiber Optic Bragg Grating Sensors", *Proc. SPIE 1169, Fiber Optic and Laser Sensors VII*, Boston, 98-107, 1989.

27. W.B. Spillman, Jr. and D.H. McMahon, "Multimode Fiber-Optic Hydrophone Based on the Photoelastic Effect", Appl. Opt. 21, 3511-3514, 1982.

28. D.H. McMachon, W.B. Spillman, Jr. and R.A. Soref, *Final Report for Naval*

Research Laboratory Contract N00014-81-C-2662, Fiber Optic Sensor System (FOSS), 1983.

29. R.E. Rudd and K.G. Goddard, *Final Report for Air Force Flight Dynamics Laboratory Contract WRDC-TR-89-3031, Composite Integrity Monitoring*, 1989.

SESSION 3

Interferometric Fiber Optic Sensors

Chair
John W. Berthold III
Babcock & Wilcox Company

The Interferometric Fiber-Optic Gyroscope

Hervé C. Lefèvre

PHOTONETICS
52, Avenue de l'Europe
78160 Marly-le-Roi - France

1- ABSTRACT

This paper reviews the technical evolution of the interferometric fiber-optic gyroscope, or I-FOG, over the last fifteen years. Today a psychological barrier has been passed, and it is now accepted that this new technology will find many applications during the 90's.

2- INTRODUCTION

Over the last fifteen years, the interferometric fiber-optic gyroscope (I-FOG) research and development has evolved from an promising physics experiment [1] to a practical device that is now close to production [2]. This has been made possible by a refined analysis of the system architecture and of the possible signal processing schemes, but also by fundamental improvements of the various optoelectronic technologies, as, in particular, integrated optics, semiconductor sources, polarization preserving fibers, and in-line fiber components.

Today, a psychological barrier has been passed, and it is now accepted among inertial guidance and control specialists that this new technology will be a strong contender for many military and civilian applications of this decade. Most leading companies in the inertial guidance field are heavily involved in R & D programs on the fiber optic gyroscopes and several have started production [2].

We are now going to describe the technical state-of-the-art and trends of present I-FOG R & D. We will also present the applications that are foreseen because of the specific advantages of the I-FOG due to its solid-state configuration : high dynamic range, high bandwidth, rapid start-up, ability to cope with a severe environment (in particular shocks and vibrations), and potential low cost.

3- BASIC ARCHITECTURE OF THE FIBER OPTIC GYROSCOPE

The interferometric fiber-optic gyroscope is a passive two-wave ring interferometer. The input light is split and propagates in opposite directions along a multiturn single mode fiber coil. Because of the Sagnac effect, the counterpropagating waves experience a phase difference $\Delta\phi_S$ when the whole system is rotated with respect to inertial space.This phase difference is proportional to the rotation rate component Ω parallel to the coil axis [3] :

$$\Delta\Phi_S = (2\pi \ L.D \ / \ \lambda.c). \ \Omega$$

where L is the length and D the diameter of the coil, λ is the source wavelength and c the light velocity in vacuo.

Practical fiber gyroscopes usually work over plus or minus half a fringe, i.e. $\pm \pi$ radian of phase difference. The theoretical sensitivity (limited by photon noise) is on the order of 10^{-6} radian or less of phase shift. Compared to the absolute phase accumulated by the waves when they propagate along the coil, this theoretical limit is extremely small, but reciprocity of light propagation in single-mode waveguides makes it detectable. This requires the use of the so-called minimum reciprocal configuration [4,5][Figure 1], which is now universally employed. This architecture needs a single spatial mode and polarization filtering at the common input-output port of the interferometer and a biasing phase modulation at one end of the fiber coil that serves as a delay line. This brings a drastic enhancement to the ring interferometer stability by making both opposite paths identical in the absence of rotation. It is also accepted that to further improve noise and drift, it is important to use a broadband source [6] and a polarization preserving fiber [7].

As a matter of fact the system comprises two primary waves sensitive to rotation but also a lot of parasitic waves (backreflected, backscattered or coupled in the crossed polarization) that yield secondary interferometers. The short coherence length of a broadband source suppresses the contrast of these spurious interferometers, and then limits their parasitic signals. The use of polarization preserving fibers is essential because such fibers are also birefringent by principle and one takes advantage of depolarization of cross-coupling with a broadband source [8,9]. This limits the required polarizer rejection to values that are compatible with present technology. Notice that white light interferometry has been found as a very powerful tool to carefully control the polarization and birefringence problems of the ring interferometer [Figure 2] [10], when one wants to get the best possible performances.

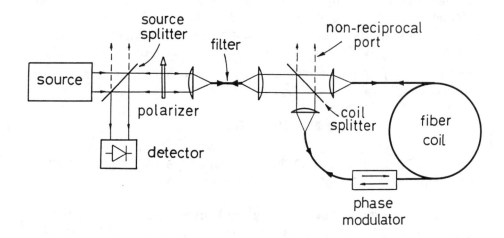

Figure 1
Minimum Configuration of the I-FOG

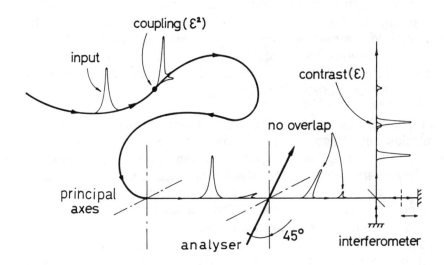

Figure 2
Test of polarization couplings
with white light interferometry

Another point of concensus is to use an all-guided technology for ruggedness. In particular it is possible to make an all-fiber gyroscope with in-line components [Figure 3] [11]. However, this approach has only been demonstrated with an open-loop signal processing which yields problems of linearity and stability of the scale factor. This can be good enough for certain applications, but most teams are now working on a closed-loop approach that allows one to linearize and stabilize the scale factor. This was first proposed with the use of bulk acousto-optic frequency shifters [12,13], but this approach yields an intrinsic source of bias unstability [14] and the general trend is now to use integrated optic modulators to implement specific closed-loop phase modulation schemes as serrodyne modulation (analog phase ramp) [15,16] or digital phase ramp [9,14,16]. Combined with a digital demodulation in an "all-digital" closed-loop processing scheme [17,18], this digital ramp feedback yields a very good scale factor linearity (10 ppm range) while preserving the good bias stability of the open-loop.

Such schemes require a broad and flat modulation bandwidth of the phase modulator that is the main technical advantage of integrated optics. Notice that the digital phase ramp could simply look like a quantified analog ramp, but it is actually fundamentally better because the digital approach is synchronized with the modulation/demodulation biasing scheme. This solves automatically the problem of the finite fly-back of the ramping by synchronous gating and it eliminates the fiber index dependence in the scale factor [14].

Integrated optics is also a favored technology for implementing several functions on the same circuit for improving compactness and simplifying the connections. In particular the optimal simplicity is obtained with the so-called "Y-tap" or "Y-coupler"configuration [Figure 4] [9,14], which uses a three-function integrated optics composed circuit of a Y junction for wavesplitting, a polarizer on the common base trunk and two phase modulators on both branches.

Such a circuit, now widely accepted and used as the " gyro circuit", has a parallelogram shape to avoid backreflections [9,14] induced by the index mismatch between the substrate and the fibers [Figure 5]. As it can be seen, the assembling of the components has a relative simplicity, requiring only a pigtailing of the source and three pigtailings on the I.O. circuit. These connections can be ruggedized to withstand a rough environment.

For sake of completeness, one should not forget two effects that are intrinsically non-reciprocal as the Sagnac effect : the magneto-optic Faraday effect [19] and the non-linear Kerr effect [20].They could have been a drastic limitation to high performance because

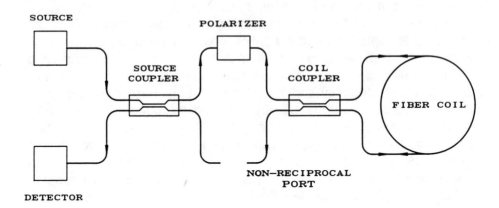

Figure 3
All-fiber gyroscope

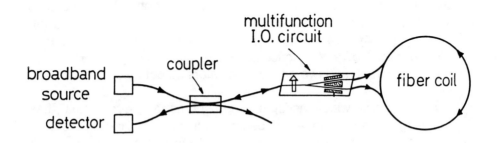

Figure 4
"Y-tap" or "Y-coupler" configuration

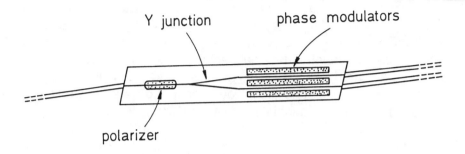

Figure 5
The "gyro circuit"

they cannot be distinguished from the Sagnac effect, but it was found that they can be eliminated very simply, and they are not a problem anymore. The use of a polarization preserving fiber eliminates the Faraday non-reciprocity if care is taken to avoid random twist in the fiber [21], and the use of a broadband source averages out the Kerr non-reciprocity [22,23]. We find here a very nice property of the interferometric FOG : there are not any parasitic effects that require incompatible solutions. On the contrary, the same solution solves nicely several different problems at the same time : polarization preserving fibers apply to the Faraday effect but also to birefringent non-reciprocity ; broadband sources eliminate backscattering and polarization coupling noise but also the Kerr non-reciprocity !

Finally thermal transients [24] and vibrations [25] can produce parasitic signals because reciprocity is strictly valid only for time invariant systems. Winding techniques must symmetrize the coil to get common mode rejection of the perturbations [24]. The so-called quadrupolar technique [26] looks like the most efficient method. Adequate potting is also necessary to define a stable pointing accuracy of the sensing coil.

4- PRESENT SUBJECTS OF CONCERN AND APPLICATIONS

FOG prototypes with very good performances have been reported by various R & D teams [2]. Their bias stability is in the 10 to 0.1 degree per hour range for compact devices (100 to 50 mm diameter) using 100 to 500 meters of polarization preserving fiber, and a superluminescent diode as a broadband source. Scale factor accuracy and linearity are in the 100 ppm range or less, for closed-loop FOG's using integrated optics and in the 1000 ppm range, for open-loop FOG's.

Present performances do fit tactical application requirements (0.1 to 10 °/h range) where they bring in addition a very high dynamic range (more than 1000 °/s) and a very large measurement bandwidth (several kHz) that cannot be reached with "classical" technologies. Interest is particularly strong for agile missiles for example. The good behaviour of FOG solid state technology under shocks and vibrations is also making it a strong candidate for smart ammunitions.

Besides military applications, the FOG looks also very promising for use in a hybrid navigation, system with a GPS (Global Positioning System) receiver. For these applications, efforts are directed towards ruggedization and reduction of cost which is expected to be very competitive once the FOG enters production. Use in cars is even seriously envisaged in Japan [2].

Advanced reseach on components is still continuing for high performance applications. It concerns superfluorescent rare earth doped fiber sources [27,28,29] that will improve the wavelength stability of the broadband source, compared to semiconductor sources. This research field is very promising for many other applications, such as in particular telecommunications, and it should help to fulfil the ppm scale factor accuracy requirement of an inertial navigation grade FOG.

Another important topic is the improvement of the rejection of the integrated optic polarizer. The usual technique is Ti-indiffused waveguides on a lithium niobate ($LiNbO_3$) substrate and polarization extinction is obtained with a metallic overlay. It is known however that proton exchange on a lithium niobate circuit yields single polarization waveguidance that gives a very high polarization extinction ratio (more than 60 dB). Great progress in the annealing technique of proton exchange has yielded gyro circuits with very attractive performances [30] that further improve the bias stability and get into the 0.01 °/h range of inertial grade navigation.

5- CONCLUSION

The fiber optic gyroscope is now widely accepted as the new gyro technology of the 90's and it will get a significant share of the market. Many companies, universities and governmental agencies are working on the subject [2] : Litton, Honeywell, Smith Industries, Bendix, MIT, Stanford University, JPL and NRL, in the USA ; Mitsubishi, JAE, Hitachi and Tokyo University in Japan ; SEL, Litef, Teldix, AEG-Telefunken, British Aerospace, Ferranti, Sagem, University College, Strathclyde University and Photonetics in Europe. There is also an important effort on components.

Various demonstration prototypes have shown very good performances and several companies are close to production. The main advantage of the FOG compared to "classical" technologies is its solid-state configuration. This yields high dynamic range, high bandwidth, rapid start-up, good resistance to shocks and vibrations, which will extend the field of applications of inertial guidance techniques even further also because of its potential low cost.

6- BIBLIOGRAPHY

• Most of the important publications about the fiber-optic gyroscope have been compiled in :
"Selected Papers on Fiber-Optic Gyroscopes",
Edited by R.B. Smith, SPIE Milestone Series, Vol. MS8, (1989).

• Two books on the fiber-optic gyroscope are going to be published by the end of 1992 or early 1993
- *"Optical Fiber Rotation Sensing"*,
Edited by W.K. Burns and to be published by Academic Press
- *"The Fiber-Optic Gyroscope"*,
by H.C. Lefèvre, to be published by Artech House.

7- REFERENCES

For references that can be found in the SPIE Milestone Series Compilation we have added : [MS 8, pages xx-yy].

[1] Vali V. and Shorthill R.W.,
"Fiber ring interferometer",
Applied Optics 15, 1099-1100, (1976), [MS 8, 134-136].

[2] Proceedings of *"Fiber Optic Gyro : 15th Anniversary Conference"*,
SPIE Proceedings, Vol. 1585, (1991).

[3] See for example : Arditty H.J. and Lefèvre H.C.,
"Sagnac effect in fiber gyroscopes",
Optics Letters, 6, 401-403, (1981), [MS 8, 105-107].

[4] Ulrich R.,
"Fiber optic rotation sensing with low drift",
Optics Letters, 5, 173-175, (1980), [MS 8, 170-172]

[5] Ezekiel S. and Arditty H.J.,
"Fiber-Optic Rotation Sensors : Tutorial Review",
Springer Verlag Series in Optical Sciences, Vol. 32,
2-26, (1981), [MS 8,   3-27]

[6] Böhm K. , Russer P. , Weidel E. and Ulbrich R.,
"Low drift fibre gyro using a superluminescent diode",
Electronics Letters, 17, 352-353, (1981), [MS 8, 181-182]

[7] Burns W.K., Moeller R.P., Villaruel C.A. and Abebe M.,
"Fiber-optic gyroscope with polarization holding fiber",
Optics Letters, 8, 540-542, (1983), [MS 8, 208-210]

[8] Fredericks R.J. and Ulrich R.,
"Phase error bounds of fibre gyro with imperfect polariser-depolariser",
Electronics Letters, 20, 330-332, (1984), [MS 8, 277-278]

[9] Lefèvre H.C., Bettini J.P., Vatoux S. and Papuchon M.,
"Progress in optical fiber gyroscopes using integrated optics",
NATO/AGARD Conference Proceedings, Vol. 383, 9A-1, 9A-13,
(1985), [MS 8, 216-227]

[1 0] Lefèvre H.C.,
"Comments about fiber optic gyroscopes",
SPIE Proceedings, Vol. 838, 86-97, (1987), [MS 8, 56-67]

[1 1] Bergh R.A., Lefèvre H.C. and Shaw H.J.,
"All single mode fiber gyroscope with long term stability",
Optics Letters, 6, 502-504, (1981), [MS 8, 178-180]

[1 2] Davis J.L. and Ezekiel S.,
*"Techniques for shot-noise-limited inertial rotation
measurement using a multiturn fiber Sagnac interferometer"*,
SPIE Proceedings, Vol. 157, 131-136, (1978), [MS 8, 138-143]

[1 3] Cahill R.F. and Udd E.,
"Phase-nulling fiber-optic gyro",
Optics Letters, 4, 93-95, (1979), [MS 8, 152-154]

[1 4] Lefèvre H.C., Vatoux S., Papuchon M. and Puech C.,
"Integrated optics : a practical solution for the fiber-optic gyroscope",
SPIE Proceedings, Vol. 719, 101-112, (1986), [MS 8, 562-573]

[1 5] Kay C.J.,
"Serrodyne modulator in a fibre-optic gyroscope",
IEE Proceedings, Vol. 132, 259-264, (1985), [MS 8, 448-453]

[1 6] Lefèvre H.C., Graindorge Ph., Arditty H.J., Vatoux S. and Papuchon M.,
"Double closed-loop hybrid fiber gyroscope using digital phase ramp",
Proceedings of OFS-3 ' 85, Post deadline paper PSP 7, (1985), [MS 8, 444-447]

[1 7] Arditty H.J., Graindorge P., Lefèvre H.C., Martin P., Morisse J. and Simonpiétri P.,
"Fiber-Optic gyroscope with all-digital closed-loop processing",
Proceedings of OFS 6'89, Paris, 131-136, (1989)

[1 8] Lefèvre H.C., Martin P., Morisse J., Simonpiétri P., Vivenot P. and Arditty H.J.,
"High-dynamic range fiber gyro with all-digital signal processing",
SPIE Proceedings, Vol. 1367, 72-80, (1990)

[1 9] Böhm K., Petermann K. and Weidel E.,
"Sensitivity of a fiber-optic gyroscope to environmental magnetic fields",
Optics Letters, 7, 180-182, (1982), [MS 8, 328-330]

[2 0] Ezekiel S., Davis J.L. and Hellwarth R.W.,
"Intensity dependent nonreciprocal phase shift in a fiberoptic gyroscope",
Springer-Verlag Series in Optical Sciences, Vol. 32, 332-336, (1981), [MS 8, 308-312]

[2 1] Hotate K. and Tabe K.,
"Drift of an optical fiber gyroscope caused by the Faraday effect : influence of the earth's magnetic field",
Applied Optics, 25, 1086-1092, (1986), [MS 8, 331-337]

[22] Bergh R.A., Culshaw B., Cutler C.C., Lefèvre H.C. and Shaw H.J.,
"Source statistics and the Kerr effect in fiber-optic gyroscope",
Optics Letters, 7, 563-565, (1982), [MS 8, 313-315]

[23] Petermann K.,
"Intensity-dependent non-reciprocal phase shift in fiber-optic gyroscopes for light sources with low coherence",
Optics Letters, 7, 623-625, (1982), [MS 8, 322-324]

[24] Schupe D.M.,
"Thermally induced non-reciprocity in the fiber-optic interferometer",
Applied Optics, 19, 654-655, (1980), [MS 8, 294-295]

[25] Lefèvre H.C., Bergh R.A., and Shaw H.J.,
"All-fiber gyroscope with inertial-navigation short-term sensitivity",
Optics Letters, 7, 454-456, (1982), [MS 8, 197-199]

[26] Frigo N.J.,
"Compensation of linear sources of non-reciprocity in Sagnac interferometers",
SPIE Proceedings, Vol. 412, 268-271, (1983), [MS 8, 302-305]

[27] Wysocki P.F., Fesler K., Liu K., Digonnet M.J.F., and Kim B.Y.,
"Spectrum thermal stability of Nd- and Er-doped fiber sources",
SPIE Proceedings, Vol. 1373, 234-245, (1990)

[28] Morkel P.R., Laming R.I., and Payne D.N.,
"Noise characteristics of high-power doped-fiber superluminescent sources",
Electronics Letters, Vol. 26, 96-98, (1990)

[29] Iwatsuki K.,
"Excess noise reduction in fiber gyroscope using broader spectrum linewidth Er-doped superfluorescent fiber laser",
IEEE Photonics Technology Letters, Vol. 3 281-283, (1991)

[30] Suchoski P.G., Findakly T.K. and Leonberger F.L.,
"$LiNbO_3$ integrated optical components for fiber optic gyroscopes",
SPIE Proceedings, Vol. 993, 240-244, (1988)

Critical review of resonator fiber optic gyroscope technology

Glen A. Sanders

Honeywell Systems & Research Center
Phoenix, AZ 85027-2708

ABSTRACT

A critical review of the developments in the resonator fiber optic rotation sensor technology is presented. Error terms and proposed solutions are discussed. Supporting experimental results and their significance are summarized.

1. INTRODUCTION

Historically, the resonator fiber-optic gyroscope[1,2] was a natural extension of earlier concepts based on a passive ring resonator technique for rotation sensing applications.[3] Similar to the interferometer fiber optic gyroscope (IFOG) and the ring laser gyroscope (RLG), the passive resonator concept utilizes the phase difference experienced between clockwise (cw) and counterclockwise (ccw) waves traveling around a closed path that is rotating with respect to an inertial frame, i.e., the Sagnac effect. In the case of the passive ring resonator, the Sagnac phase shift gives rise to a resonance frequency difference between counterpropagating light waves in a optical ring resonator. While earlier efforts focused on mirrored optical resonators,[4] the advent of low loss single mode fiber-optic rings,[5] offered moderately high quality resonators with the ability to achieve longer length multi-turn resonators, and potentially low-cost rugged implementations. Because of this, the resonator fiber optic gyro (RFOG) has received the most attention amongst the various possible optical resonator configurations.

The basic RFOG concept is shown conceptually in figure 1. In this technique, light of intensity "I" and tunable frequency "f" is introduced into a fiber resonator in the ccw direction. The resonator consists of a short loop of fiber and a fiber-optic coupler. As the

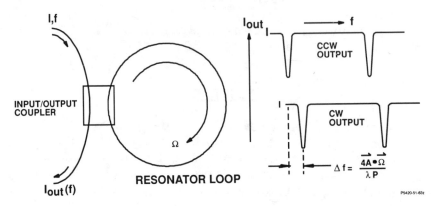

Figure 1. RFOG principle of operation.

frequency is tuned such that an integral number of wavelengths fit inside the optical pathlength of the ring, the input energy is absorbed into the recirculating loop and a sharp resonance dip is shown on the output. As light is injected in the opposite direction (cw), similar resonances are observed. In the presence of rotation the resonance frequencies are split by a frequency splitting Δf, given by

$$\Delta f = \left| \frac{D}{\lambda n} \right| \Omega \tag{1}$$

where λ is the free space wavelength of the light, n is the index of refraction of the fiber, and D is the diameter of the loop assuming a circular coil shape. A typical RFOG configuration[6] is shown in figure 2. In this case (but not all cases), light is tapped out of the ring by use of a second coupler giving resonance peaks in the "transmission mode" operation of the resonator rather than the "reflection mode" shown in figure 1. The resonances are tracked by servo loops,[3,4,7] one that adjusts the laser frequency f_0 to the ccw resonance center and the other that adjusts a frequency shift Δf to allow resonance tracking in the cw direction. Each servo is based on sinusoidal frequency modulation (at f_n or f_m) of the input light waves and subsequent demodulation of the output to accurately determine resonance centers. The frequency Δf is then a measure of rotation rate given by equation (1).

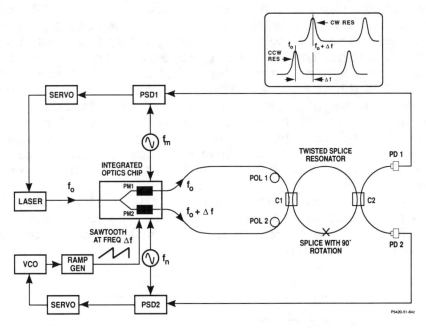

Figure 2. One RFOG configuration investigated at Honeywell.

The resonator fiber-optic gyro has been pursued for reasons similar to that of the IFOG; namely it's all-solid-state construction offers the potential for very high reliability and long life, fast turn-on time, and light weight. Its additional potential to go one step beyond the IFOG by achieving the same shot-noise-limited performance with a factor of 10-100x less fiber[8] has attracted various developers. This is a direct consequence of the finesse of the ring which quantifiably signifies the fact that light traverses the loop multiple times as opposed to once in the IFOG counterpart. Figure 3 illustrates the fundamental sensitivity advantage over the IFOG for the same fiber length. The change in intensity for a given optical pathlength change is shown for both cases. The RFOG finesse, given by the separation between peaks divided by the full width half maximum intensity determines its maximum slope (sensitivity). Finesses of up above 100 are quite practical. The IFOG has an equivalent finesse of 2. Because of this, the uncertainty in the measurement of optical pathlength in the presence of noise is approximately F/2 lower in the RFOG case. Alternatively, it should be possible to trade this extra sensitivity for reduced fiber length. This feature gives the RFOG a significant potential advantage over the IFOG for high performance applications where the coil length of the IFOG (approximately 1 km) becomes a significant portion of the device cost and size. Another advantage of the shorter coil was pointed out by Shupe,[1] who showed that bias sensitivities due to thermal non-reciprocities should scale with fiber length, thus favoring the resonant device. In fact, it is possible for short fiber resonator rings to employ single layer winding schemes that allow direct isothermal contact of the whole coil to a conducting bobbin.

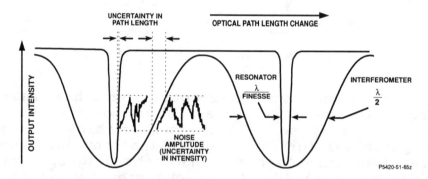

Figure 3. Fundamental sensitivity comparison for RFOGs and IFOGs assuming same detected power and sensing length.

The key goal of RFOG research is thus to reduce other sources of noise and drift below that of the shot noise to take advantage of the RFOG's intrinsic high sensitivity. While progress has been made toward achieving moderate performance levels, the RFOG performance feasibility remains the primary challenge for the <0.1 deg/hr where envisioned RFOGs are expected to be most attractive relative to other technologies. Encouraging progress continues as a result of research into the key error mechanisms and the demonstration of solutions. This progress is the primary focus of this review. The key error mechanisms discussed include optical backscatter, the optical Kerr Effect and polarization-type errors. A review of the impact of these effects on gyro performance, as well as data illustrating the efficacy of various solutions, is included in this paper.

2. PERFORMANCE ISSUES, SOLUTIONS AND DEMONSTRATED RESULTS

Perhaps the greatest technical challenge for RFOG developers has been the need for precise determination of the cw an ccw resonance frequencies of the ring to accurately determine the rotation rate. In fact, to reach optimum (shot-noise-limited) performance a typical measurement accuracy of less than 10^{-7} of the linewidth is necessary. In addition to the use of AC detection techniques to precisely measure the resonance center, the resonance lineshape itself must be highly symmetric. Optical scattering from the cw to the ccw direction (and vice versa), as well as the excitation and detection of a second state of polarization in the ring compromise this lineshape symmetry. In contrast, the optical Kerr Effect actually produces a resonance splitting similar to that of rotation. In this section, each of these effects is discussed.

2.1. OPTICAL BACKSCATTER EFFECTS

Perhaps one of the most universal issues amongst optical gyroscopes is that of optical backscattering which compromises the desired independence between the counterpropagating waves. These effects are virtually eliminated by the uses of mechanical dither for the RLG and the broadband light source for the IFOG. In this section, a brief discussion of the backscatter issue in the RFOG is included as well as a review of the methods developed for overcoming these performance limitations.

Suppose light waves of amplitudes E_{CCW} and E_{CW} and frequencies f_1 and f_2 are incident on the fiber ring in the ccw and cw directions, respectively. It is assumed for simplicity of illustration (shown in figure 4) that the optical backscatter within the ring is due to a single point scatterer with amplitude coefficient η (the argument may be generalized to the case of distributed Rayleigh backscatter). The field detected at the cw output is then a combination of the main signal wave E^{out} at frequency f_2 as well as a backscattered ^{CW} wave E_s at frequency f_1, each of which have resonance characteristics. The primary wave has experienced a round trip phase delay inside the ring that includes ϕ, the cw rotation-induced phase shift; whereas the backscattered wave is a more complicated function of both ϕ and $-\phi$ due to its propagation in both directions. The output intensity in the cw direction is given by

$$I_{cw} = \left| E^{out}_{cw}(f_2,\phi) \right|^2 + 2\mathrm{Re}\left[E^{out}_{cw}(f_2,\phi) E^*_s(f_1,\phi,-\phi) \right]$$

$$+ \left| E_s(f_1,-\phi,\phi) \right|^2 \qquad (2)$$

The first term is the primary signal, which has a "resonance dip" characteristic every time f_2 is tuned to resonance in the cw direction (ideal case). The middle term in equation 2 is the interference term which is proportional to the backscatter coefficient η and is the product of two fields at different frequencies. Not only does this term give intensity variations at $f_1 - f_2$ on the detector, but it also gives the appearance of a cw resonance frequency that oscillates at $f_1 - f_2$, i.e. "beats." This is illustrated in figure 5 for a typical RFOG configuration[4] where both waves are detected at the same modulation frequency, fm. Assuming that the ccw output is used for matching the laser frequency to the ccw resonance frequency, the open-loop output of the cw phase sensitive demodulator will be a measure of rotation rate. For low rotation rates, the beats at $f_1 - f_2$ fall well inside the bandwidth of the detection electronics, and an oscillatory bias is observed. Typical error amplitudes are on the order of a few degrees per second[9] due to Rayleigh backscatter within the ring.

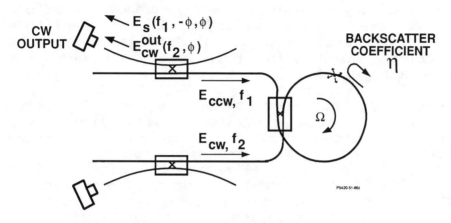

Figure 4. Optical backscatter in the RFOG.

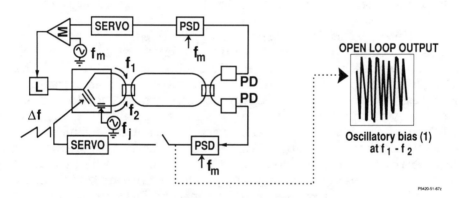

Figure 5. RFOG arrangement using a common modulation frequency f_m for CW and CCW resonance detection, and the observed oscillatory bias due to the backscatter interference term.

In a closed-loop system, i.e. where this oscillatory error signal is allowed to feed a servo that tracks the second resonance via frequency shifting, it has been shown[10] that "lock-in" behavior occurs. In this case, the indicated frequency output of the gyro, which should be a linear function of the rotation rate, shows a "dead zone" at low rotation rates. This is very similar to the behavior observed in a non-dithered ring laser gyro.

The most basic of the proposed[4] solutions to this error term was the use of sinusoidal phase or frequency modulation (at frequency f_j) of one input wave as shown in figure 5. The amplitude of this modulation is adjusted so as to null the carrier which, in principle, changes the frequency of the error to positive and negative integer multiples of f_j plus $f_1 - f_2$. Assuming f_j is outside the bandwidth of the detection/servo electronics these error signals are filtered out, as shown in the middle of figure 6 for the open-loop gyro case. This technique was also shown[10] by Ezekiel et al to eliminate the corresponding lock-in effect observed in closed-loop operation. The error suppression, however, is a sensitive function of the modulation drive amplitude which determines the degree of carrier suppression. Error reductions on the order of 100x are feasible in practice using this basic solution.

Several other variations and enhancements of this error-averaging technique have been proposed and demonstrated. One of these methods for further reducing this error utilizes pseudo random noise[11] in addition to the phase modulation at f_j which further randomizes the output and eliminates the residual oscillations. This was demonstrated in our laboratory and is shown on the right-hand side of figure 6. While this scheme is quite effective, its drawback is that it introduces excess random noise (above that of the shot noise) in the gyro output. Other variations include the application of carrier suppressed phase modulations to both input waves[9] or actually using the carrier suppression in one input but implementing the resonance detection in the cw and ccw directions at different frequencies. This latter technique was suggested by Iwatsuki et al[12] and demonstrated in our laboratories to suppress errors below the 0.4°/hr resolution of our most recent breadboard gyro.

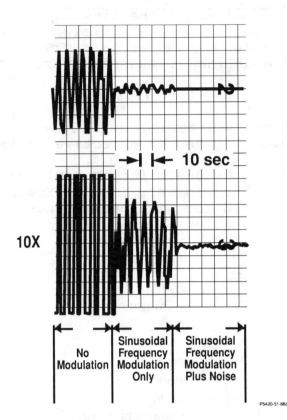

Figure 6. Error due to the influence of backscatter in the RFOG and illustration of proposed solutions. Left-hand side of top plot is the oscillatory error at $f_1 - f_2$. The middle plot is the residual error after application of approximately carrier suppressed sinusoidal modulation. The right-hand portion shows the error after random noise is added in addition to the sinusoidal modulation. Bottom curves are the same with 10 magnification.

The third term in equation 2 is the intensity of the backscattered wave. Since this term has resonance lineshape characteristics that depend on both propagation directions, i.e. a dependence on $-\phi$ and $+\phi$, this signal component at the detector has two peaks. The first peak location (see figure 7) corresponds to the resonance center in the cw direction and the second peak location is at the ccw resonance frequency of the ring. In the absence of rotation the cw and ccw resonance frequencies are the same, causing the lineshape centered about the cw resonance frequency to have a single peak. In this case, it contributes no error. As the rotation rate increases, the lineshape splitting grows, contributing an error proportional to rate, i.e. scale factor error.[12] The resultant scale factor non-linearity can be approximated by

$$\epsilon = 2 \eta^2 (F/\pi)^2$$

The Rayleigh backscattering in a 30 meter ring at 1.3 μm wavelength gives an effective backscatter coefficient of $\eta^2 \approx 4.5 \times 10^{-6}$. Thus, for a finesse of 100, the scale factor nonlinearity is roughly 0.9%, which is well in the excess of the 30 - 50 ppm needed for high performance inertial navigation applications.

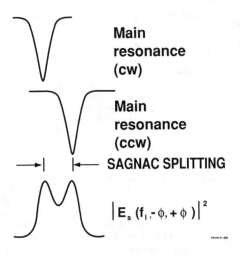

Figure 7. Lineshape of main resonances and backscatter intensity term under the influence of rotation.

This issue is easily eliminated by use of separate modulation and resonance detection frequencies as mentioned above and shown in figure 2. The ccw and cw inputs are thus modulated at frequencies f_m and f_n, with the corresponding photodetector outputs demodulated at frequencies f_m and f_n respectively. Because the scattered field arriving at the cw output was derived from the ccw input and is not modulated at f_n, this term is effectively ignored by the signal processing. A similar argument holds for the ccw output.

We have investigated the use of separate modulation frequencies in combination with carrier suppressed modulation at a third frequency in one input to the ring. Our results are shown in figures 8 and 9. In the data shown in figure 8, a production GG1342 ring laser gyro (RLG) (known to have better than 10 ppm scale factor accuracy) and the RFOG were mounted on a rate table and rotated back and forth by hand. The data shows that the RFOG output tracks that of the RLG quite well. No dead zones occur near low rotation rate, signifying that the backscatter interference term has been suppressed. To further quantify the scale factor performance, more tests were taken[7] and the results are shown in figure 9. The scale factor nonlinearity was 860 ppm or less, indicating a minimum measurable suppression of the backscatter intensity-type error term of at least a factor of four for this setup.

4.   OPTICAL KERR EFFECT

The sensitivity of the fiber gyroscope to an optical power mismatch between cw and ccw waves, was first observed by S. Ezekiel, J. Davis and R. Hellwarth for the IFOG case.[13] This effect is consequence of the dependence of the propagation constants β_{cw} and β_{ccw} on the intensities of the light waves in the non-linear fiber medium. In the RFOG, a power imbalance translates into an optical pathlength difference in the ring, causing a resonance frequency splitting similar to that of rotation. As shown in figure 10, this power imbalance may be caused by an unequal split ratio at the input ($U_1 \neq U_2$), or by a difference in modulation depth ($\phi_1 \neq \phi_2$) between the two phase modulation signals used to sense the resonance[14] center in each direction. In the latter case, unequal modulation excursions result in different time-averaged intensities in the ring. By simplifying the relation in reference 14, the rotation-equivalent error $\Delta\Omega$ is approximately given by,

$$\Delta\Omega \approx I_0 b \left[\left(\frac{F}{\pi} \right) (U_1 - U_2) + \sqrt{2} \left(\frac{F}{\pi} \right)^2 \Delta\phi \right]$$

$$(3)$$

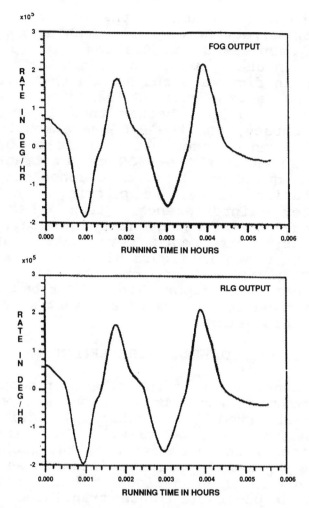

Figure 8. The simultaneous outputs of both the RFOG and a reference ring-laser gyro which were rotated together on the same table for comparison.

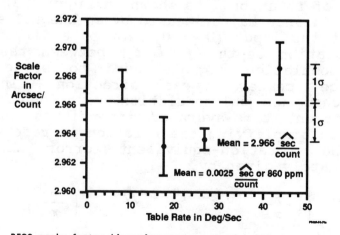

Figure 9. RFOG scale factor linearity versus rotation rate.

where F is the finesse of the ring, I_o is the source intensity, U_1 and U_2 are the fractional power magnitudes in the ccw and cw inputs to the ring, and

$$b = \frac{c \eta n_2}{aA} \qquad (4)$$

where c is the speed of light in a vacuum, n_2 the Kerr coefficient, η the impedance of the fiber, a is the radius of the fiber coil and A is the effective area where the light is focussed inside the fiber. As shown in equation 3, the error scales with the quantity F/π, since this factor represents the power amplification ratio between light inside ring and the input. The higher the finesse "F" the greater the power inside the ring, which magnifies the power imbalance and enhances $\Delta\Omega$. As an example, a configuration similar to that shown in figure 10 is assumed with $I_o = 10\mu W$. This represents the estimated minimum power necessary to achieve a shot noise limited performance consistent with 0.01°/hr navigation grade performance. Assuming $U_1 = 0.49$, $U_2 = 0.51$, F=100, a wavelength of 1.3μm, a 10 cm coil diameter, $\Delta\phi = 0$, and the calculated value of n_2 from reference 13, then $\Delta\Omega = 3.8°/hr$. The error is thus significantly larger than the desired shot noise limited performance goal.

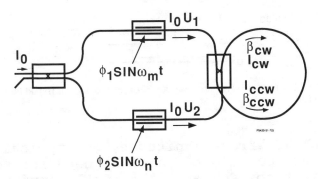

Figure 10. Simplified RFOG schematic showing sources of Kerr-Effect errors.

The second term in equation 3 is due to the power imbalance created by unequal modulation amplitudes. In this case $\Delta\phi = \phi_1 f_m - \phi_2 f_n$ represents the difference in round trip phase shift inside the ring produced by unequal external phase modulation amplitudes ϕ_1 and ϕ_2. This difference is a direct consequence of using separate modulation frequencies for cw and ccw waves to prevent previously discussed backscatter-induced scale factor errors. This error is more sensitive to the finesse (than that associated with $U_1 - U_2$), since a higher-finesse lineshape produces more intensity variation at the output for a fixed modulation excursion; hence, a greater power imbalance for a given $\Delta\phi$. Although this second term may be larger than the first term in equation 3, it may be more stable since variations in the electro-optic coefficient of the integrated modulators are common to both inputs, and modulation drive voltages may be controlled.

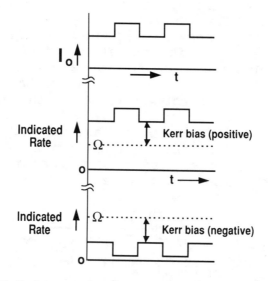

11. Method for sensing optical Kerr-Effect bias during RFOG operation.

The above error magnitudes, for typical RFOG design cases, point out a serious need for active compensation to achieve the full performance potential of this device. One scheme which was proposed[15] and has been demonstrated by Takiguchi et al,[16] looks quite promising. This countermeasure, illustrated in figure 11, makes use of the fact that the Kerr-induced bias may actually be measured during gyro operation. As shown in figure 11 (also see figure 12), the light source intensity is modulated at frequency f_K (top of figure 11). If there is no Kerr-bias, i.e. $\Delta\phi = 0$ and $U_1 - U_2 = 0$, then there will be no subsequent bias variation at f_K. In the presence of a Kerr-induced error, the gyro indicated rate will vary proportionately and coherently at f_K, with its phase depending on the sign of the bracketed term in equation 3. If the sign of this term is positive, the gyro output variations will be in phase with the intensity modulation (middle of figure 11); likewise, if the sign is negative, the indicated output at f_K will be $180°$ out of phase with the intensity modulation (bottom of figure 11). These results are independent of the real rotation rate or direction. As shown in figure 12, the gyro output may be demodulated at f_K to obtain an error signal, which in turn, may be used to null out the Kerr bias. This may be accomplished by tuning an intensity transducer, located in one gyro input path, to correct for power imbalances between cw and ccw waves in the ring.

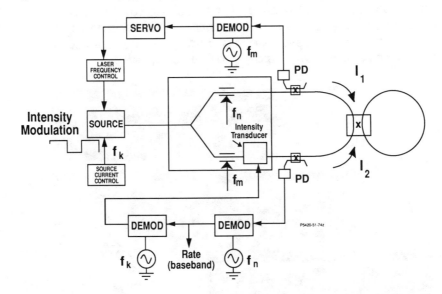

Figure 12. Schematic for nulling of Kerr-Effect errors in the RFOG.

5. POLARIZATION ISSUES

Another important RFOG issue stems from the existence of a second polarization state in the ring.[2,17] The propagation of light in this second state is supported by the fiber and most often excited by cross-coupling in the fiber optic coupler or by imperfect launching conditions at the input to the ring. The existence of this undesired light in the ring gives rise to a second resonance dip which causes gyro bias errors. In a single mode ring the location and size of the second dip is a sensitive function of environmental parameters since the birefringence of the fiber varies with external conditions. In a polarization-maintaining ring, which we have investigated in our laboratory, the birefringence is a predictable but rapid function of temperature.[18] This causes the behavior shown in figure 13. While the resonances are separated, the intended resonance (for rotation sensing) and the unwanted resonance nominally correspond to light propagating along the principle axes of the fiber (x- and y- polarized). As the temperature changes, the birefringence changes rapidly causing the differential round trip pathlength between polarizations to change. For a typical resonator ring, the two dips coincide about every one degree celsius. During this condition the center of the lineshape is indiscernible, causing gyro errors on the order of 1 deg/sec.

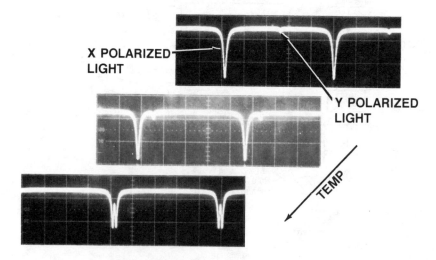

Figure 13. Output of fiber ring resonator having resonance dips for the two polarization states. The ring was constructed using PM fiber, hence the resonances nominally correspond to x- and y-linearly polarized states while separated. As the birefringence changes with temperature, the resonances drift together.

The first solution proposed was the use of single polarization fiber for the ring[2] or the use of a polarizer in the ring. By introducing loss on the y-axis of the fiber for instance, the magnitude of the second dip is substantially reduced; in fact, for relatively modest polarizer extinction ratios the second dip is virtually eliminated. Dahlgren et al[19] has demonstrated one such ring with relatively low loss (finesse of 25) using a 40dB-extinction- ratio polarizer. In this case, only a single resonance dip was observed.

There are remaining issues, however, for the single polarization ring. One of these polarization-related error mechanisms[20,21] is illustrated in figures 14 and 15. A typical single polarization ring is shown, with linear polarizers of extinction ratio ϵ located on its input leads. In practice it must be assumed that the incident light contains the unwanted polarization state (E_y^i) as well as the main signal input E_x^i. Assuming the undesired light is attenuated in amplitude by ϵ in the first polarizer, it is then incident on the resonator.

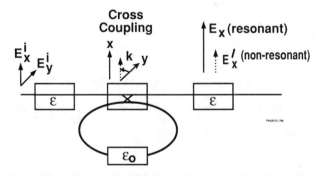

Figure 14. Cause of polarization error in single polarization ring due to imperfect launching conditions and components. The undesired, non-resonant light interferes with primary resonant wave.

Assuming it is off resonance, it travels through the throughput port of the coupler where it is partially cross-coupled to the x-axis. The cross-coupling amplitude coefficient is represented by k. This undesired light of amplitude E_x then passes through the second polarizer unattenuated where it can then interfere with the primary signal. The intended input E_x^i is also incident to the ring, and assuming the gyro is operating, is near a resonance condition. This is represented by the output dip shown in figure 15. The phase ϕ_x of the intended output signal varies rapidly through resonance as shown in figure 15, while the non-resonant light has constant magnitude ($|E_x'| = |E_y^i|\epsilon k$) and phase.

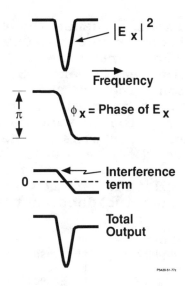

Figure 15. Magnitude and phase of primary light, field amplitude of undesired light and resulting interference signal as a function of optical frequency. Bottom trace shows the lineshape asymmetry observed at the photodetector for the case shown in figure 14.

The interference term between the two output fields thus changes sign on either side of the main resonance, as shown in figure 15. Hence, the total output intensity approximated by

$$ |E_x|^2 + 2|E_x||E_x'|\cos(\phi_x - \text{const}) \qquad (5) $$

appears as an asymmetric resonance.[20] Assuming roughly a -30dB input launching condition ($|E_y^i|^2$), $\epsilon^2 = -60$dB, $k' = -30$dB, a finesse of roughly 100 and a resonator length of 20 - 30m, the resulting error is in the 1-5°/hr range.

In summary, single polarization designs have been proposed and demonstrated as a means of eliminating the largest polarization effects due to excitation of the second-polarization resonance dip. While residual polarization errors remain, modest performance is now possible based on available component performance.

A second solution to the polarization issue has been proposed[22,23] and demonstrated in our laboratory. This approach uses polarization maintaining fiber throughout the loop. In order to overcome the temperature phenomena shown in figure 13, a 90° rotation of the polarization axes of the fiber is employed inside the ring. This may be accomplished either by means of a splice, as shown in figure 16 or by rotating the fiber axes in the resonator coupler. This causes the light in the ring to propagate equal distances along the two principal axes of the fiber, thus canceling the effects of the fiber birefringence.

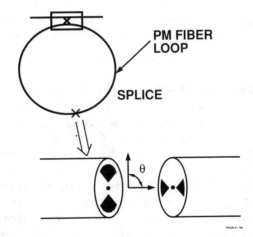

Figure 16. Polarization-rotating fiber resonator.

The two orthogonal resonant waves are thus right circular and left circular, 45° linear and -45° linear, etc., respectively, as they propagate along the length of the ring. The result of forcing each resonant wave to propagate along both fiber axes is that the two resonance dips do not cross each other with temperature; in fact, they are separated by half the free spectral range as shown in figure 17, independent of temperature. If the splice angle deviates slightly from 90°, the second resonance peak will oscillate about the half-free-spectral range position with increasing temperature, but for easily achievable splice accuracies the resonances never cross.

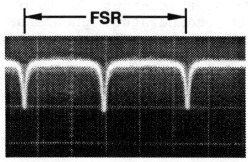

P5420-51-79z

Figure 17. Output of PR resonator showing separation of one half free spectral range between resonances of two resonant polarization states.

We have constructed several polarization-rotating resonators, the latest of which comprised two fused couplers and a 25m long sensing coil, as shown in figure 18a. By detecting at the output of the second coupler, the resonator is used in a "transmission" mode; where resonance peaks rather than dips are observed (figure 18b). Use of the low loss fused coupler technology developed at Honeywell provided for a resonator finesse of 50 over the entire -40°C to +95°C temperature range (figure 18c), thus achieving the first environmentally stable high finesse ring.[24] In fact, with an input power of 10μW, and a diameter of 3 inches, this device has the intrinsic sensitivity (shot-noise limit) necessary to achieve a navigation grade random noise performance of 0.003°/rt-hr.

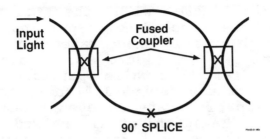

Figure 18a. Two-coupler, transmission mode polarization rotating ring resonator.

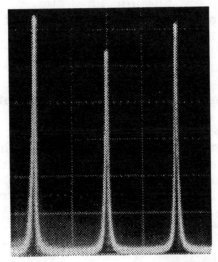

P5420-51-81z

Figure 18b. Resonances peaks of fiber resonator shown in 18a.

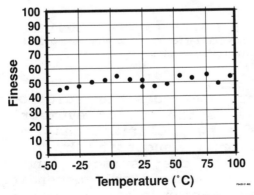

Figure 18c. Resonator finesse versus temperature for ring configuration shown in figure 18a implemented with fused couplers.

The main issue with the polarization rotating (PR) ring according to our analysis[6] is shown, although greatly exaggerated for the sake of illustration, in figure 19. This effect stems from the fact that while operating the gyro at resonance for say "state 1," there is a non-zero excitation of "state 2" even though it is off-resonance. Takiguchi et al[25] showed that the intensity of that second resonance contributes errors sufficiently small enough to allow the achievement of navigation grade performance. This agrees with our calculations. However in practice an interference term has to be considered. Even though the two fields may be orthogonal in principle, and do not interfere at the output, this condition may be spoiled by unequal losses for the two waves. Two possible examples are unequal coupling ratios or unequal insertion losses in the output coupler. This causes the coupler to combine and interfere two waves that may otherwise be orthogonal, analogous to two orthogonal linearly polarized waves passing through an analyzer at 45°. The result is lineshape asymmetry at the output as shown in figure 19. This asymmetry can cause gyro biases on the order of 30°/hr for practical components.

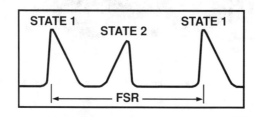

Figure 19. Illustration of apparent resonance lineshape asymmetry in polarization-rotating ring resonator for two resonant polarization states.

There are two possible solutions to this problem. One is the use of symmetric couplers that treat the two polarizations equally. The second solution is suggested by our analysis, which shows that the asymmetry and asymmetry-induced error reverses itself depending on which resonant polarization state is being detected. Alternately switching between resonances thus affords a method for averaging this effect.[26]

A gyro utilizing these principles was assembled and tested in our laboratory. It consisted of 50m PM fiber ring in a transmission mode (see figure 20a) with a 90° splice. The diameter of the ring was 6" with a finesse of 30. A 1.3µm Nd:YAG laser source was used. The gyro performance data was quite encouraging (Figure 20b) showing an output bias stability of the gyro of close to 0.4°/hr over a two-hour period. This data[6] was actually shown to be limited by our rather high random noise of 0.1 deg/rt-hr which is due to imperfect laser noise cancellation when using separate modulation frequencies for cw and ccw resonance detection. Lasers with lower noise and the application of refined laser stabilization techniques are now being investigated.

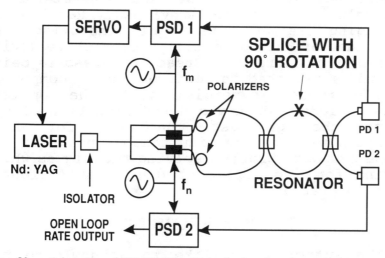

Figure 20a. Schematic of RFOG built with polarization rotating ring.

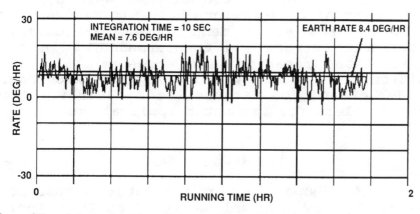

Figure 20b. Output stability of gyro configuration shown in figure 20a.

In summary, the PR resonator approach solves the most important polarization problem by spectrally isolating the second resonance from the intended resonance. Considerable progress has also been made towards identification of residual errors and solutions. Further performance enhancements are expected with improved components and signal processing techniques.

6. SUMMARY

The status of RFOG has been reviewed with particular attention to the fundamental issues of optical Kerr effect, optical backscatter and polarization effects. Several solutions to these problems have been proposed and the status of these solutions have been reviewed. Results are encouraging as new ways of improving the performance of this potentially attractive device are being demonstrated. Steady progress is being made and is expected to continue due to component and electronics synergism with the optical communications industry and the IFOG technology. Perhaps the greatest challenge for RFOG developers will be the simultaneous usage of the various proposed problem-solving techniques while striving for a design that can be sufficiently simple in practice.

7. ACKNOWLEDGEMENT

The author would like to acknowledge the work of Lee K. Strandjord of Honeywell Systems and Research Center for helpful discussions regarding RFOG error mechanisms, particularly polarization errors and countermeasures in the polarization rotating ring resonator. In addition, the author would like to acknowledge Sam Habbel and Joe Straceski for their work on fused couplers for resonator applications.

8. REFERENCES

1. D. M. Shupe, "Fiber resonator gyroscope: sensitivity and thermal nonreciprocity," Appl. Optics, 15 Jan. 1981, Vol. 20, No. 2, pp 286-289.

2. R. E. Meyer, S. Ezekiel, D. W. Stone, and V. J. Tekippe, "Passive fiber-optic ring resonator for rotation sensing," Optics Letts., Dec. 1983, Vol. 8, No. 12, pp. 644-646.

3. S. Ezekiel and S. R. Balsamo, "Passive ring resonator laser gyroscope," Appl. Phys. Letts., 1 May 1977, Vol. 30, No. 9, pp 478-480.

4. G. A. Sanders, M. G. Prentiss, and S. Ezekiel, "Passive ring resonator method for sensitive inertial rotation measurements in geophysics and relativity," Opt. Letts., Nov. 1981, Vol. 6, No. 11, pp 569-571.

5. L. F. Stokes, M. Chodorow, and H. J. Shaw, "All single-mode fiber resonator," Opt. Letts., June 1982, Vol. 7, No. 6, pp 288-290.

6. L. K. Strandjord and G. A. Sanders, "Resonator fiber-optic gyro employing a polarization-rotating resonator," Proc. SPIE Vol. 1585, Fiber Optic Gyros: 15th Anniversary Conference, 1991, pp 163-172.

7. G. A. Sanders, G. F. Rouse, L. K. Strandjord, N. A. Demma, K. A. Miesel, and Q. Y. Chen, "Resonator fiber-optic gyro using $LiNbO_3$ integrated optics at 1.5 μm wavelength," Proc. SPIE Vol. 985, Fiber Optic and Laser Sensors VI. (Conf., 6-7 Sept. 1988, Boston, Mass.), SPIE, Bellingham, 1988, pp 202-210.

8. S. Ezekiel, H. J. Arditty, <u>Springer Series in Optical Sciences, Fiber-Optic Rotation Sensors and Related Technologies</u>, Proc. of the First International Conf. MIT, Cambridge, Mass., Nov. 9-11, 1981, - Springer- Verlag Berlin Heidelberg New York 1982, Tutorial Review, pp 2-26.

9. T. J. Kaiser, D. Cardarelli, and J. Walsh, "Experimental developments in the RFOG," Proc. SPIE Vol. 1367, Fiber Optic and Laser Sensors VIII (1990), pp 121-126.

10. F. Zarinetchi and S. Ezekiel, "Observation of lock-in behavior in a passive resonator

gyroscope," Opt. Letts., June 1986, Vol. 11, No. 6, pp 401-403.

11. Private communication with J. E. Killpatrick and N. Demma of Honeywell Systems and Research Center, January 1986.

12. K. Iwatsuki, K. Hotate, and M. Higashiguchi, "Effect of Rayleigh backscattering in an optical passive ring-resonator gyro, Appl. Optics, 1 Nov. 1984, Vol. 23, No. 21, pp 3916-3924.

13. S. Ezekiel, J. L. Davis, and R. W. Hellwarth, "Intensity Dependent Nonreciprocal Phase Shift in a Fiberoptic Gyroscope," <u>Springer Series in Optical Sciences, Fiber-Optic Rotation Sensors and Related Technologies</u>, Proc. of the First International Conf. MIT, Cambridge, Mass., Nov. 9-11, 1981, - Springer- Verlag Berlin Heidelberg New York 1982, pp 332-336.

14. K. Iwatsuki, K. Hotate, and M. Higashiguchi, "Kerr effect in an optical passive ring-resonator gyro," J. Lightwave Tech., June 1986, Vol. LT-4, No. 6, pp 645-651.

15. U.S. Patent 4,673,293.

16. K. Takiguchi and K. Hotate, "Method to reduce the optical Kerr-effect induced bias in an optical passive ring resonator gyro," Proc. of 8th Optical Fiber Sensors Conference, Monterey, CA, Jan. 1992, pp 34-37.

17. K. Iwatsuki, K. Hotate, and M. Higashiguchi, "Eigenstate of polarization in a fiber ring resonator and its effect in an optical passive ring-resonator gyro," Appl. Optics, 1 Aug. 1986, Vol. 25, No. 15, pp 2606-2612.

18. G. A. Sanders, N. Demma, G. F.Rouse, and R. B. Smith, "Evaluation of polarization-maintaining fiber resonator for rotation sensing applications," Optical Fiber Sensors (OFS'88) (5th Intl. Conf., 27-29 Jan. 1988, New Orleans); 1988 Tech. Digest Series, Vol. 2, pt. 2, Opt. Soc. Am., Wash. D.C., pp 409-412.

19. R. P. Dahlgren and R. E. Sutherland, "Single polarization fiber-optic resonator for gyro applications, Proc. of the SPIE, Vol. 1585, 1991, pp 128-135.

20. W. Schroder, P. Grollman, J. Herth, M. Kemmler, K. Kempf, G. Neumann and S. Oster, "Progress in fiber gyro development for attitude and heading reference systems," Proc. of the SPIE, Vol. 719 Fiber Optic Gyros: 10th Anniversary Conference, 1986, pp 162-168.

21. R. Carroll, "Polarization Crosstalk in a PM fiber resonator gyro," Proc. of the SPIE, Vol. 1169, Fiber Optic and Laser Sensors VII, 1989, pp 388-399.

22. G. A. Sanders, R. B. Smith and G. F. Rouse, "Novel polarization-rotating fiber resonator for rotation sensing applications," Proc. of the SPIE, Vol. 1169, Fiber Optic and Laser Sensors VII, 1989.

23. P. Mouroulis, "Polarization fading effects in polarization-preserving fiber ring resonators," Proc. of the SPIE, Vol. 1169, Fiber and Lasers Sensors VII, 1989, pp 400-412.

24. B. Kelley, G. A. Sanders, C. E. Laskoskie, and L. K. Strandjord, "Novel Fiber-Optic Gyroscopes for KEW Applications," presented at AIAA Aerospace Design Conference, Irvine, CA, Feb. 3-6, 1992.

25. K. Takiguchi and K. Hotate, "Evaluation of the output error in an optical passive ring-resonator gyro with a 90-deg polarization-axis rotation in the polarization-maintaining fiber resonator," IEEE Proton Tech. Lett., Vol. 3, No. 1, Jan. 1991, pp 88-90.

26. L. K. Strandjord and G. A. Sanders, "Performance improvements of a polarization-rotating resonator fiber-optic gyroscope," presented at the SPIE Conference on Laser and Fiber Sensors X, Boston, MA, Sept., 1992.

SESSION 4

Distributed and Multiplexed
Fiber Sensors

Chair
Juichi Noda
NTT International Corporation (Japan)

DISTRIBUTED OPTICAL FIBER SENSORS

John P Dakin

Optoelectronics Research Centre
University of Southampton
Southampton SO9 5NH, U.K.

ABSTRACT

This critical review paper covers the field of distributed optical fiber sensors, where measurements may be taken along the length of a continuous section of optical fiber. Such a feature greatly increases the information that can be obtained from a single instrument and hence the cost per sensing point can be more acceptable.

The review will not attempt to cover all methods, but will give a selection of some of the more interesting theoretical concepts, describe the current status of research and indicate where optical sensing methods are being applied in commercial instruments.

1. INTRODUCTION

The highest state of the art in optical sensing is achieved with optical fiber distributed sensors. Such sensors permit the measurement of a desired parameter as a function of length along the fiber. This is clearly of particular advantage for applications such as "smart" skins, as a sensor can measure the variation of, for example, temperature over significant areas of the outer layer of vehicles.

There are three main criteria which must be satisfied to achieve a distributed sensor. Firstly, it is necessary to construct (or select) a fiber which will modify the propagation of light in a way which can be relied upon to be dependent on the parameter to be measured. Secondly, one must be able to detect the changes in transmission (or light scattering) arising from the parameter to be measured. Thirdly, it is necessary to locate the region of the fiber where the change in propagation occurs, in order to achieve the desired spatial distribution. The paper will commence with a discussion of the basic methods which can be employed. These methods will then be expanded upon in later sections, and a number of examples of promising and practical sensors will be described. The range of applications is very large, but, as mentioned above, there is currently a strong interest in sensors for smart structures. Potential applications in this area will therefore be given some emphasis, although a broader presentation of the technology will be given.

2. BASIC CONCEPTS OF DISTRIBUTED SENSORS

A distributed sensor consists of a continuous length of fiber, usually with no taps or branches along its length. It is therefore necessary to determine the location of any measurand-induced change in transmission or scattering properties by taking advantage of the propagation delays of light travelling in the fiber. Such propagation delays allow differences in the time of arrival of light, travelling in different modes of propagation, to be related to distances along the fiber.

The greatest differences in propagation delay occur when signals are travelling in opposite directions in the fiber. The simplest example of this is the optical time domain reflectometer (OTDR), where a pulsed signal is transmitted into one end of the fiber, and returning back-scattered signals are recovered from the same fiber end, (figure 1).

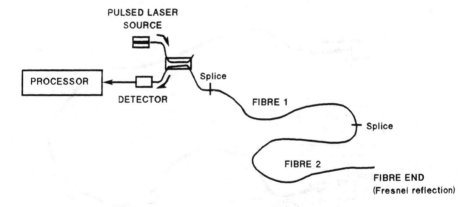

(a) BASIC OPTICAL ARRANGEMENT OF OPTICAL TIME DOMAIN REFLECTOMETER (OTDR)

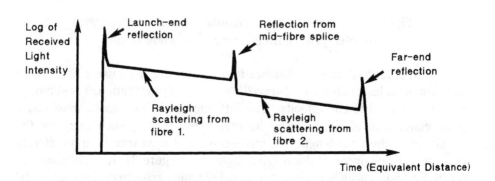

(b) INTENSITY VERSUS TIME, OTDR RETURN.

Figure 1   Concept of the basic optical time domain reflectometer

The concept is a guided-wave optical variant of the radar-location principle, where distance, z, is related to the two-way propagation delay, 2t, by the simple formula:-

$$z = t.V_g \tag{1}$$

where V_g is the group velocity of light in the fiber. Rather than rely on weak backscattered light, it is possible to use counter propagating light, and arrange for a non-linear optical interaction to occur when the beams cross. Thus a continuous "probe" beam can be modulated by a "pump" pulse travelling in the opposite direction (see figure 2).

Changes in the probe intensity with time can again be related to changes with position along the fiber. A third method of location involves signals which travel only in the forward direction.

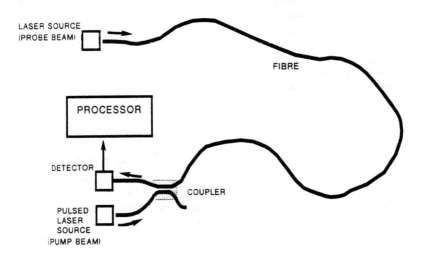

**Figure 2   Schematic of distributed sensor concept with
counter-propagating pump and probe beam**

In order to determine distance in this case, one must use a fiber which will support at least two forward-travelling modes of propagation. These modes must travel at different velocity. The influence to be monitored must cause some conversion of light energy between these modes, such that, for the remainder of the fiber length, the mode-converted light travels in a different mode to the remainder of the original light, (see figure 3). It is necessary to detect the small changes in propagation delay which arise from the intermodal difference in propagation velocity over this latter section. These changes are so small that it is generally necessary to use some form of interferometric approach in order to detect them.

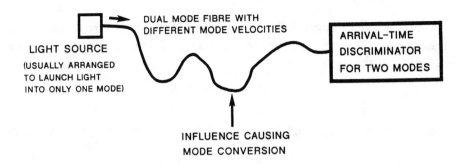

**Figure 3   Basic concept of distributed sensing using
fowalr-travelling waves only**

Having outlined the basic methods, each will now be described in more detail and variants of each of the basic concepts will be discussed.

3. BACKSCATTERING METHODS, INCLUDING OTDR

3.1 General concept

The OTDR, first reported in reference 1, has been established as a standard item of fiber optic test gear for many years. Its main application is for fault finding and attenuation monitoring in optical networks. As described above, the OTDR relies on backscattering (or back reflection) of light which has been launched into a fiber from an amplitude-modulated (usually-pulsed) source. The light is Rayleigh scattered from refractive index fluctuations in the core of the fiber, or may be reflected from discontinuities,such as connectors, splices, fiber breaks etc. The Rayleigh scattering component, which is detected returning from the arrangement shown in figure 1, is given by the following equation (reference 2) :-

$$P(z) = \tfrac{1}{2}\, S(z).\, \alpha_s(z).\, V_g.\, \exp\left\{ -\int_o^{z'} [\alpha_f(z') + \alpha_b(z')]\, dz' \right\} \qquad (2)$$

where P(z) is the detected backscattered power, as a function of the distance, z, of the scattering point along the fiber, S(z) is the captured fraction of scattered light coupled into backward-travelling modes in the fiber, α_s (z) is the scattering coefficient of the fiber, V_g is the group velocity of light in the fiber, and α_f and α_b are the total attenuation coefficients of the fiber in the forward and backward directions, respectively. (Usually α_f and α_b will have the same value; i.e:- $\alpha_f = \alpha_b = \alpha$). As stated in the introduction, the distance, z, is related to the two-way time of flight, 2t, by the relation:-

$$z = t.V_g$$

When the attenuation coefficients and the capture factor are constant, the expression for the detected backscattered power, P, has an exponential time dependence of the form:-

$$P(t) = A_1 . \exp(-B_1.t) \tag{3}$$

where A_1 and B_1 are constant. This is, of course, the situation for a uniform fiber.

The OTDR can sense changes in the total attenuation coefficient, α, if the scattering coefficient, α_s, and the capture fraction, S, are constant. Under these conditions:-

$$P(z) = A_2 \cdot \exp\left[-\int_0^z \alpha(z').dz'\right] \tag{4}$$

where A_2 is a constant. The rate of change (differential with respect to time) of the detected signal is proportional to the attenuation coefficient. Alternatively, it can sense changes in scattering coefficient α_s if α and S are constant:-

$$P(z) = A_3 . \alpha_s(z).\exp(-B_3 z) \tag{5}$$

where A_3 and B_3 are constants. (Clearly α will normally change with changes in α_s, but if the value of the integral $\int_0^z \alpha(z').dz'$ is small, the error will be small).

Many parameters can cause (or be arranged to cause) variations in the attenuation or scattering coefficient of a fiber. These will be discussed in the following section.

3.2 Distributed Sensors Based on Monitoring Attenuation Variations with the OTDR

3.2.1 Microbend Sensors for Mechanical Sensing

Any distortion of an optical fiber from its ideal cylindrical shape gives rise to an increase in attenuation. Gradual bends, having a radius above a few cms, generally result in a much smaller loss than very small and sharp bends. Highly localised bends, or kinks, of this type are known as microbends, and cause significant losses in both monomode and multimode fiber. Generally, in monomode fiber, the loss due to a given bend is fairly amenable to theoretical analysis (reference 3). For this type of fiber, it is only necessary to determine the loss of energy from the fundamental mode. Thus, monomode fibers are potentially capable of quantitative determination of the extent of bending of a fiber simply from a measurement of attenuation, provided that the shape of the bend deformation is known.

Multimode fibers, on the other hand, are less predictable in their behaviour. A particular problem is that the attenuation due to a given bend is a complex function of the manner in which power is distributed between the multiplicity of waveguide modes at the entrance to the bend. This mode excitation is a function not just of the launching conditions, but also of the entire topology of the fiber (including any other microbend sensors!) prior to the measurement region. Thus, multimode microbend sensors will generally have some potential value for use as simple qualitative sensors, but are less likely to make practical quantitative types.

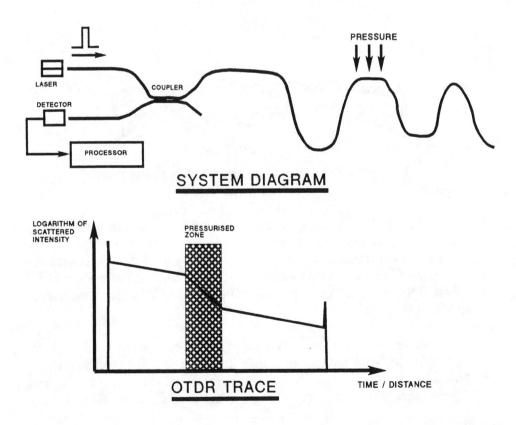

Figure 4   Concept of distributed sensing using a
pressure-sensitive cable

In order to construct a microbend sensing system, it is necessary to arrange for the parameter to be measured to cause microbending of a cable which is monitored with an OTDR system (figure 4). A convenient form of sensing cable is one of a type originally designed by Harmer, which is now a commercial product (Herga Ltd., reference 4; see also figure 5). This cable contains an inner communications fiber, with a polymer fiber wound spirally round it. These fibers are then sheathed within a close-fitting outer tube. When compressed, the outer tube squashes against the spiral polymer fiber, which in turn deforms the inner optical fiber to give it a periodic, alternating, lateral displacement. Multimode fibers suffer particularly high microbending loss if the spatial period of the distortion matches the pitch of the 'zig-zag' path taken by the highest order modes in the straight fiber.

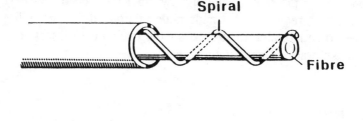

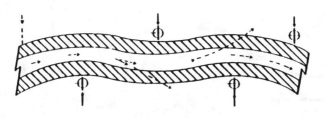

Figure 5   Principle of the microbend-sensing fiber cable (Hergalite)

The spiral-wound fiber cable forms a distributed sensing element which is capable of many qualitative sensing tasks. For example, it can sense the pressure of a footstep and can hence form a security barrier for safety reasons (for example, to switch off a dangerous machine or other manufacturing process if a person approaches: figure 6) or to protect against unauthorised intruders (figure 7).

There are many sensing systems of this type, where it is sufficient for a simple transmission monitor to merely detect the total line-averaged effects of disturbance along a single section. However, the cable is capable of location of disturbance along a continuous length if it is coupled to an OTDR system.

**Figure 6   A pressure-sensing mat for machine guarding
(Herga Electric Ltd., U.K.)**

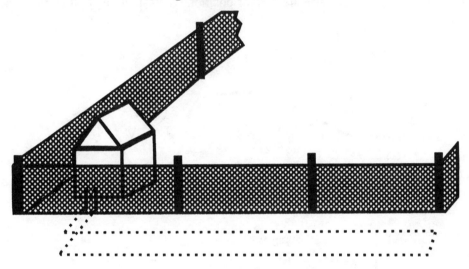

**Figure 7   Schematic of application of fibre-optic intruder
detector system buried in ground**

In addition to detection of footpressure, the microbend sensor has been considered as a sensor for the monitoring of civil engineering structures (reference 5). Clearly, if it can be made in a reliable form, it has considerable potential for use as a strain sensor for "Smart Structures". If the strain in a structure can be reliably converted to a periodic ripple in a fiber, then the structure can be qualitatively monitored. However, if multimode fiber is used in the sensor, then the performance is likely to be rather variable, because the response will be dependent on the mode conversion occurring in earlier sections. In addition, the polymer elements which transfer the strain to the inner fiber can creep at room temperature and will flow more readily at high temperatures. Much work remains to be done, therefore, to develop versions suitable for high temperatures, such as may arise in aerospace applications.

3.2.2 Radiation Sensing

The attenuation of an optical fiber increases when exposed to ionising radiation. In order to cause attenuation, the radiation must be able to penetrate to the core region of the fiber, and must have sufficient energy to cause structural change in the glass core. Three types of radiation affect attenuation:- neutrons, γ rays and X-rays. (The latter, being also photons, are essentially the same as γ rays in nature, but they differ in the way in which they are generated and are usually of lower energy). Gaebler and Braunig (reference 6) were the first workers to recognise the potential of an optical fiber for distributed radiation sensing and to demonstrate the method experimentally using an OTDR. The basic concept of distributed radiation sensing is shown in figure 8. The irradiated section of fiber has a higher loss, and hence the gradient of the OTDR trace changes for positions corresponding to this region.

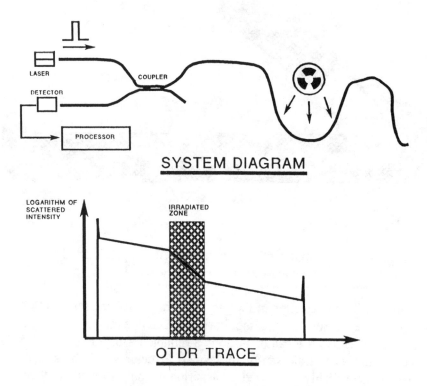

**Figure 8 Concept of the distributed radiation dosimeter
(W. Gaebler and D. Braunig 1983)**

The distributed radiation sensor has particular attractions for detecting the presence of a single localised area of irradiation, as the line-integrated loss is reasonably low and the probe signal is not unduly attenuated. The sensor reported in reference (6) was probably less well suited for quantitative radiation dosimetry, as the sensitivity of normal germania-doped silica fiber is poor, and the response can exhibit a strong dose-rate dependence. The measurement can suffer significant error due to room temperature annealing which will slowly reduce the induced loss.

3.2.3 Temperature Sensing

The first distributed temperature sensor, by Hartog and Payne, was based on a liquid-filled fibre (reference 7, figure 9). This fiber is simply a silica glass tube filled with a higher refractive index, low-absorption, liquid, which acts as the light-guiding core of the waveguide. The scattering loss coefficient, α_s, of the liquid depends on the density fluctuations caused by thermodynamic molecular motion, and therefore shows a strong temperature dependence.

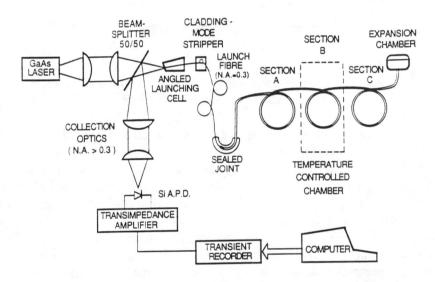

Figure 9   Rayleigh-scattering temperature profiler using liquid-filled fibre (reference 7)

The thermal variations are not significant in glass fibers, as the scattering is caused by "frozen-in" density fluctuations, formed as the glass was cooled from the melt. The scattering variations in the liquid-filled fiber are directly observable from an OTDR trace, (figure 10), due to the dependence of the return signal on the scattering loss coefficient, α_s, in section 3.1. This sensor successfully demonstrated the ability to measure temperature distribution, for the first time, but the use of a liquid-filled fiber renders it impractical for a wide variety of applications. We shall therefore consider a means of using a solid glass fiber with an OTDR system. (Other methods using glass fiber will be discussed in later sections).

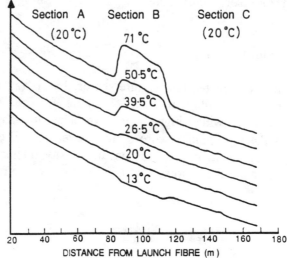

Figure 10   Results of distributed temperature sensing using OTDR in liquid-filled fibre

Optical fibers doped with rare-earth ions, such as Nd^{3+} and Ho^{3+}, show strong absorption peaks in the visible and near infrared regions of the spectrum. Many of these absorbing transitions occur between electronic levels having a temperature-dependent occupancy, and therefore several absorption peaks have a significant thermal dependence (ie the coefficient α is temperature dependent). The OTDR can probe the absorption as a function of length, and hence determine the variation of temperature along the fiber (reference 8). As already mentioned, a disadvantage of using variations in an absorbing fiber to sense a parameter is that both the probe and return back scatter signals are rapidly attenuated by this same absorption. The method is nonetheless attractive for systems requiring only a few measurement elements along the length of the fiber. It is also an excellent method when a fiber is used which has a low attenuation in normal use, yet which increases if sections become either too hot or too cold. This is clearly useful for methods requiring the detection and location of hot-spots or cold-spots, particularly if these are likely to occur only over a short length of the fiber.

The absorption loss along the length of a holmium-doped fiber, as measured at 665 nm, is shown in figure 11 (reference 9). This fiber was cooled to -196°C in liquid nitrogen for much of its length, but the central region was allowed to increase in temperature in stages, being eventually heated to +40°C and +90°C. The attenuation profile shows the potential value of such a sensor for detecting loss of cryogenic coolant. It could, for example, also be used to detect the effects of solar heating on an otherwise-cold spacecraft surface. Many other combinations of rare earth dopants, host fiber and interrogation wavelengths are possible, and there may therefore prove to be suitable candidates for sensors for the detection of a wide range of over- and under-heating and cooling conditions.

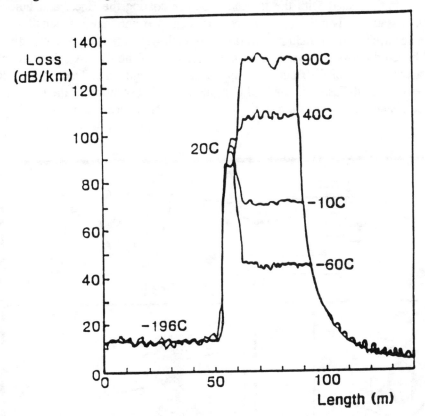

Figure 11 **Attenuation versus length in an Ho^{3+}-doped fiber, when a central section is exposed to varying temperature (outer sections are held at -196°C). Loss was measured with an OTDR at 665nm (reference 9)**

In addition to fibers using special dopants, there is one type of commonly available fiber which has a significant temperature coefficient of attenuation. This is the polymer-clad-silica (PCS) fiber using a silicone cladding. This fiber has a low attenuation around room temperature and above, but suffers severe attenuation when cooled to below -50°C or so. This effect has been exploited as a practical sensor for detecting leakage in cryogenic liquids which has been launched as a commercial product, (reference 10). The attenuation occurs due to increase in the refractive index of the cooled cladding, which leads to eventual loss of guidance when it approaches the index of the silica core.

3.2.4 Chemical Sensing Using Absorbing Coatings

Distributed chemical sensors can be constructed by coating an optical fiber with indicator chemicals. As mentioned, a pure silica fiber can be coated with polymers of lower refractive index, such as silicone resins. These resins can be impregnated with the indicator, before coating the fiber, and before polymerisation. The guided modes in such a fiber have a small, yet significant,evanescent field penetration into the lower index polymer cladding, and hence the loss is dependent on the absorption of the dye. A schematic of the arrangement for distributed sensing is shown in figure 12. The chemical to be sensed can diffuse into the cladding, modify the absorption of the dye, and hence change the attenuation of the fiber. The first experimental results for

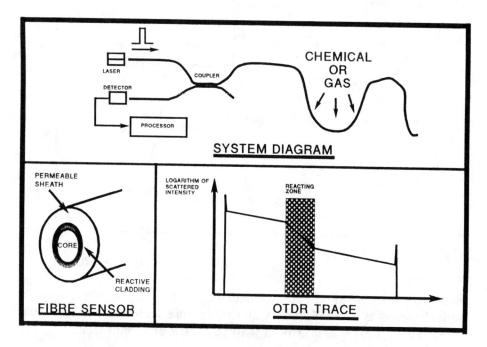

Figure 12   Concept of the distributed chemical sensor

such a sensor were obtained by performing a simple transmission measurement on a single section of fiber. Results for ammonia detection, using the change in colour of a pH indicator, were reported (reference 11). It is quite possible that, by the time this review goes to print, there may be reports of distributed chemical sensors capable of resolving the location of the interactions.

The evanescent-field chemical sensor has a number of potential problems which need to be solved before reliable operation can be achieved. The main problems arise from the variations in evanescent field coupling due either to changes in the temperature of, or adsorption of chemicals (including water) in, the cladding material. In addition, any small bends or fluctuations in the diameter of the fiber, will modify the evanescent field coupling. Finally, there is a need for care to ensure a sufficiently rapid diffusion time to ensure a fast enough response, and also to choose materials which will display a reasonably reversible behaviour, yet avoid any problems of the indicator becoming leached out.

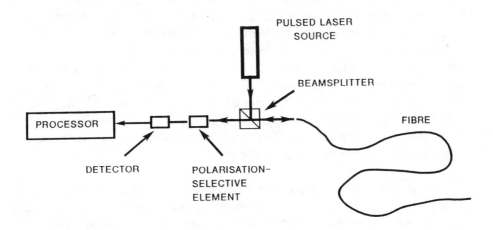

Figure 13   Basic optical arrangment of polarisation optical time domain reflectometer (POTDR) (reference 13)

Even if sophisticated distributed chemical sensors cannot be perfected, they may find application in a much simpler form in smart structures. It has already been shown possible to detect oil leaks, using the increase in cladding refractive index when it becomes contaminated with the oil, (reference 12). This causes severe loss in the fiber, and could therefore form a means of detecting leaks of fuel, lubricant or hydraulic oil. For "smart skins" applications, this will be useful to detect leaks onto structures from containment vessels or transport piping.

3.3 Polarisation Sensing (Polarisation Optical Time Domain Reflectometry or POTDR)

The POTDR method (reference 13; figure 13) is similar to OTDR, except that a pulse of polarised light is launched and the detector is arranged to be polarisation sensitive by placing a polariser before it. The method relies on the fact that Rayleigh and Rayleigh-Gans scattering in silica glass is polarised in the same direction as the incident light. Thus, any changes in the polarisation of the detected light result from changes during propagation over the two-way path to and from the scattering point. Changes in polarisation of light from different points along the fiber resulting differences in the detected signals at the appropriate delay times.

The POTDR method was first suggested by Rogers (reference 13), who pointed out its potential for distributed measurements of magnetic field (via Faraday rotation), electric field (via the Kerr quadratic electro-optic effect), lateral pressure (via the elasto-optic effect) and temperature (via the temperature dependence of the elasto-optic effect). The first experimental measurements were reported by Hartog et al. (reference 14), who used the technique for distributed measurement of the intrinsic birefringence of a monomode fiber. Kim and Choi (reference 15) measured the birefringence induced by the bending of a wound fiber. Ross (reference 16) carried out the first measurement of a variable external field using the Faraday rotation of polarisation caused by the magnetic field environment of the fiber. A comprehensive theoretical treatment of the POTDR method has been presented in the specialist paper of Rogers (reference 17).

The POTDR technique appears to be attractive for the measurement of a large number of parameters. However, as with many potentially useful sensing methods, its main drawback is the variety of parameters to which it can respond. Spurious sensitivity to strain and vibration are particularly troublesome. In addition, POTDR requires the use of monomode fibers, which can, when used with narrow linewidth laser sources, give rise to particular problems from coherent addition of light returning from multiple Rayleigh backscattering centres (reference 18). Significant recent research is being directed towards solving the vibration problems as the current monitor has strong commercial potential. More recent POTDR work has been carried out on birefringent fiber. The beat-length variations of such fiber are temperature dependent. Temperature distribution can be determined by monitoring the frequency of fluctuations of the detected signal when light is launched into both modes and the returning signal is passed through a polariser, (reference 19). In order to reduce the frequency of the fluctuations to a convenient range, a relatively low birefringent fiber is necessary.

3.4 Enhancement of the OTDR Using Active Fiber Components

We shall now consider how the performance of the basic OTDR can be enhanced, using a variety of recently developed active fiber components.

3.4.1 Enhancement of the OTDR using a Fiber Amplifier

An optical fiber amplifier can be inserted within the optical circuit of the OTDR, as shown in figure 14. The position of the amplifier is such that both the outgoing pulse from the laser source, and the weak backscattered signal from the fiber, are amplified in the same device. This is capable of

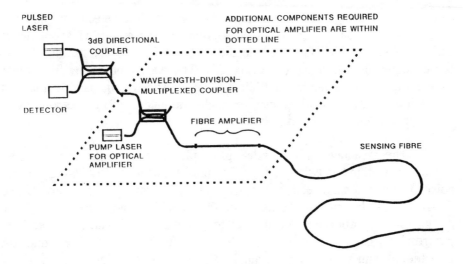

Figure 14 Enhanced OTDR system using an optical amplfier
(reference 9)

having a dramatic effect on the performance of the system. At the time of writing, such systems have not yet found significant use in sensing systems. However, their potential is considerable, (reference 9), justifying their inclusion as an important part of this review.

The outgoing laser pulse power could, in principle, be amplified at least 1000 times in a typical erbium fiber amplifier. (This has typically >30dB gain at 1500 nm, and recent laboratory results have shown up to 50dB). In a low-duty-cycle pulsed mode of operation, the small signal gain can be approached. Thus, for example, a semiconductor laser source of 20 mW power can be boosted to launch 10 Watts into the fiber, even after allowing for 3 dB loss in the 3 dB coupler. For short laser pulses of higher input power, much higher output powers of several hundred Watts are possible from the amplifier. The peak output power for low-duty-cycle short pulses can be very high, as a result of the high energy storage possible in the device. In practice, the maximum power output is more likely to be limited by other considerations, such as the onset of stimulated Raman scattering, (particularly if monomode fibers are measured), and additional constraints such as limitation of eye hazard.

The effect on the returning backscattered signal is also highly significant. Firstly, the amplifier provides gain before the lossy 3 dB directional coupler, and hence can provide an immediate 3 dB optical power improvement in the detectable signal level. However, for high bandwidth signals, the performance of the detector, in combination with an optical fiber preamplifier, has generally a far superior detection limit than with the detector alone. This advantage becomes particularly marked if the detector is equipped with a narrow band filter, included to remove any amplified spontaneous emission (ASE) which falls outside the pass band necessary to amplify the fast-changing optical signal (reference 20). The ASE suppressing filter must have a broad enough bandwidth to include any spurious frequency fluctuations in the source laser (including also the modulation sidebands corresponding to the desired spatial resolution to be covered).

The advantages of the optical preamplifier are most marked when the required detection bandwidth is high. In high bandwidth receivers, the thermal noise of conventional detection systems tends to rise rapidly, particularly above 200 MHz bandwidth. In addition, for the optical preamplifier system, the design of an appropriate optical filter to remove out-of-band ASE becomes progressively easier. An example of the detection limit improvement that can be achieved with an erbium fiber amplifier, for a received signal bandwidth of 1 GHz, is given in the paper by Laming et al, reference 21. At 1 GHz, an improvement of 7 dB in receiver sensitivity is possible. In an OTDR system, when the 3 dB gain due to amplification before the 3 dB coupler is included, a total potential receiver sensitivity improvement of 10 dB is anticipated. At frequencies above 1 GHz, as might be necessary to achieve ultra-high resolution measurements (a few cms), the potential improvement factor is much greater.

If one multiplies all the potential improvement factors available with an amplifier of moderate (30dB) gain, there is a possible improvement factor of up to 10,000 times (40dB), permitting a fiber sensor to operate with an additional 20 dB one way loss. However, care must be taken to avoid problems of spontaneous oscillation due to optical feedback (caused, for example, by reflections from the source laser and detector, or from coherent Rayleigh backscatter in the sensing fiber). The maximum possible improvement may therefore be less than this in practice. In addition, care will have to be taken to ensure that the gain of the amplifier does not change significantly during the period when the backscatter signal is returning, (or, if it does, to compensate for gain changes that occur). Provided the launched pulse does not seriously deplete the excited-state population in the amplifier (by virtue of its short duration), there is not likely to be a serious population change problem with the much weaker backscattered signals, as even the time-averaged signal power will be much less than that of the launched laser signal.

3.4.2 Use of High-power Q-switched Fiber Laser Sources

High-peak-power, short-duration, pulsed sources are required in a number of distributed sensors, in order to provide good spatial resolution and signal-to-noise ratios. Q-switched neodymium-doped fiber lasers are attractive sources for this application, by virtue of their ability to produce short duration pulses using relatively low power laser diode pump sources. Neodymium-doped silica fibers can be fabricated to give 0.3-0.5 dB gain (at 1.06 μm for each mW of launched pump power (at 810 nm). Such pump light can easily be obtained from AlGaAs laser diodes. The use of high concentration (> 1000 ppm) Nd^{3+} doped fiber permits short fiber lengths to be used, thus minimising the photon lifetime in the cavity. This is an important requirement for the generation of short-duration Q-switched pulses.

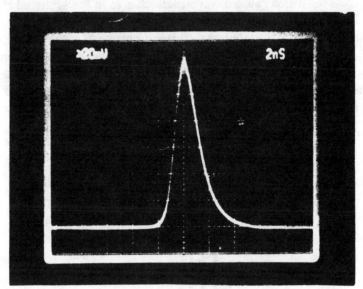

**Figure 15   A pulse from a Q-switched Nd^{3+} fiber laser
(peak power 600W, duration 5 nsec)**

A typical cavity configuration for a 1.06 μm Nd^{3+}-doped fiber laser would include a mirror with a high reflectivity at 1.06 μm, a doped fiber (length < 10cm), an intra-cavity lens, an acousto-optic or electro-optic modulator, and a partially-transmitting output coupling mirror. Pump light at 810 nm from a laser diode is launched through the input-end mirror, which has a high reflectivity at 1.06 μm, but good transmission at 810 nm. Figure 15 shows a typical Q-switched Nd fiber laser output pulse, with a peak power of ~600 W and a pulse duration of 5 ns. Such pulse characteristics would enable a spatial resolution of 0.5 m to be obtained in a distributed sensor system. For these reported results, a 100 mW laser diode, (SDL5411), was used as the pump source. For Nd^{3+} doped fibers, the pulse characteristics remain unchanged up to a repetition rate of 400 Hz, above which the peak power reduces and the pulse duration increases. It should be emphasised that Q-switched fiber laser sources are more than just a research component, having already been incorporated into a commercial distributed temperature sensor (reference 22).

3.5 Distributed temperature sensing using Raman scattering, Brillouin scattering and fluorescence

All the backscattering methods described so far have relied on elastic scattering processes. These are ones where the scattered light is at the same wavelength as the incident light. Rayleigh scattering and Fresnel reflections are both examples of elastic scattering. There are several commonly occurring physical processes which cause a wavelength change in the scattered light. In quantum theory, this implies either a loss or gain in energy by an incident photon, when creating a scattered one. (It is this change of energy that explains the use of the term inelastic, by analogy with energy loss in inelastic mechanical systems). A spectral plot of several inelastic processes is shown in figure 16. The use of these for distributed sensors will be described in the following sub-sections.

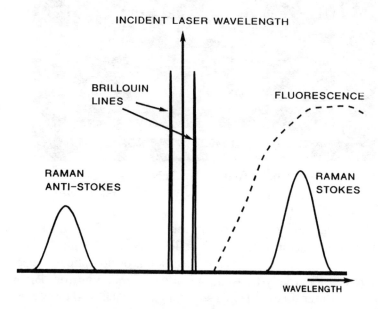

INCIDENT LASER WAVELENGTH

BRILLOUIN LINES

FLUORESCENCE

RAMAN ANTI-STOKES

RAMAN STOKES

WAVELENGTH

Figure 16   Inelastic scattering phenomena   used for sensing

3.5.1 Distributed Temperature Sensing using Spontaneous Raman Scattering

If the spectral variation of backscattering from a germania-doped silica fiber (reference 23) is examined (figure 17), it may be seen that there is a strong central line, primarily due to elastic Rayleigh (or Rayleigh-Gans) scattering. This central line also contains lines due to Brillouin scattering, which are much closer in frequency to the incident frequency, and hence more difficult to resolve. At each side of the central line, however, there are side-lobes due to Raman scattering, which are shifted in frequency to a much greater degree. These Raman lines may be used to detect temperature profiles in conventional (vitreous) communications fiber, using an OTDR system with additional filters to select out the Raman light (reference 24). From standard texts on Raman scattering, the ratio, R, of anti-Stokes (higher frequency band) intensity to Stokes (lower frequency band) intensity at wavelengths $\lambda_{a\text{-}s}$ and λ_s, respectively, (separated at equal frequency shift, $\Delta\nu$, from the central line) is given by the relationship:-

$$R(T) = \left(\frac{\lambda_s}{\lambda_{a\text{-}s}}\right)^4 \exp - \left(\frac{h.c.\Delta\nu}{K.T}\right) \tag{6}$$

where h is Planck's constant, c is the velocity of light in vacuo, k is Boltzmann's constant and T is the absolute temperature.

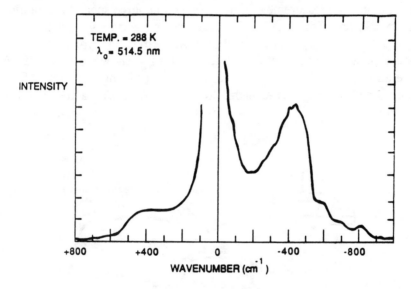

Figure 17   Raman scattering spectrum from a conventional germania-doped silica fibre

Thus, a measurement of the ratio of Stokes and anti-Stokes backscattered light in a fiber should provide an absolute indication of the temperature of the medium, irrespective of the light intensity, the launch conditions, the fiber geometry and even the material composition of the fiber. In practice, however, a small correction has to be made for variations in fiber attenuation between the different Stokes and anti-Stokes wavelengths. The Raman technique appears to have only one significant practical drawback; that of a relatively weak return signal. Usually, the anti-Stokes Raman-scattered signal is between 20 and 30 dB weaker than the Rayleigh signal. In order to avoid an excessive signal averaging time, measurements have been taken using pulsed lasers which are capable of providing a relatively high launched power. In the first experimental demonstration of the method (reference 24), a pulsed argon-ion laser was used in conjunction with Corning, telecommunication grade, 50/125 μm GRIN fiber.

The experimental arrangement is similar to that of the conventional OTDR, except that a wavelength-selective directional coupler is used, which allows the launch laser to be coupled into the fiber, but directs backscattered light at Raman wavelengths onto different detectors to receive Stokes and anti-Stokes bands, respectively.

More recent experimental results (references 25, 26) and first generation commercially developed systems (reference 27) have used a pulsed semiconductor laser and avalanche photodiode detectors, greatly reducing the size and power requirements (see figure 18). Such first generation systems have been capable of measurement of up to several hundred separate resolution elements over a few Kms of fiber. A measurement accuracy of better than $\pm 1°C$ is possible with resolution of a few metres. A typical form of temperature versus distance plot from a Raman DTS is shown in figure 19. A later generation system (reference 22) has an option of using Q-switched fiber laser sources and the range is extended to over 10 Km, with a resolution of around 1 metre.

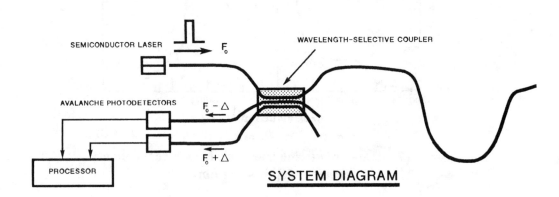

Figure 18   System diagram for Raman distributed temperature sensor, using semiconductor source and detector

The Raman DTS is now well-established as a practical sensor and has a low cross-sensitivity to other parameters such as strain, pressure, variations in type of fiber and cable, etc. It has applications for monitoring electrical machines, cables and transformers, location of fires, and for sensing industrial plant. It is clearly attractive as a temperature sensor for smart skins, being capable of measuring temperature at many thousands of points with a single instrument and fiber cable.

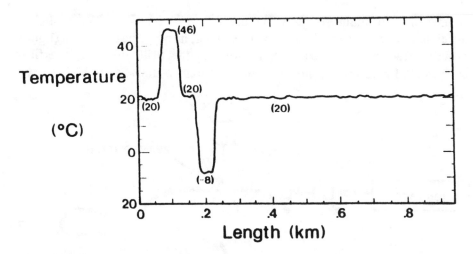

**Figure 19 Typical temperature versus distance display
for Raman DTS system**

3.5.2 Sensing using Brillouin Scattering

In addition to Rayleigh scattering and Raman scattering, glasses also exhibit Brillouin scattering. In a classical model, this latter form of scattering can be considered to be diffraction of light from the refractive index variations arising from acoustic waves. The light can be considered to be Doppler shifted as a result of the movement of the acoustic wave. The Doppler shift is characteristic of the acoustic velocity in the glass, which is a function of both temperature and pressure. In the more accurate quantum model, the light is more correctly considered to be scattered as a result of particle interactions, the incident photons being scattered from acoustic phonons, with an appropriate energy change in the scattered photon. This energy change in the photon represents a frequency change equivalent to the Doppler shift expected from the less-correct classical model. The frequency shift with Brillouin scattering is very small (≈ 12GHz) so, unlike Raman scattering, it is difficult to select out Brillouin lines with conventional optical filters. However, the intensity of Brillouin scattering is at least an order of magnitude higher than that of Raman signals, so it is an attractive possibility for sensing. There have been various research attempts to detect these lines, all by mixing (or heterodyning) optical signals scattered from the fiber with light which is frequency shifted from the original laser source. [ie., the source used to excite the Brillouin scattering] (references 28, 29; figure 20). The frequency-shifted light, required as a reference for the mixer, could be conveniently derived using a Brillouin fiber laser, with the original laser as a pump source. (A Brillouin fiber laser is simply a fiber with reflective end-mirrors, optically pumped at a level sufficient to cause lasing action via the phenomena of stimulated Brillouin scattering). This laser light is shifted, essentially to the same extent as the spontaneous Brillouin scattering in the measurement fiber. As a slightly different frequency is required for heterodyning, it should be tuned to a slightly different frequency, for example, by forming the Brillouin laser in a fiber composed of a different material to the measurement fiber.

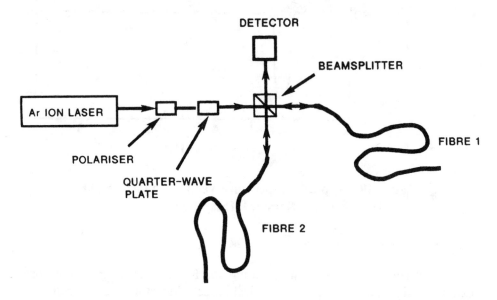

Figure 20 Basic optical arrangement for mixing of stimulated Brillouin scattering signals to create a heterodyne or difference frequency (reference 28)

The potential attraction of Brillouin scattering as a means of sensing is that the response to the measurand, such as a change in temperature or strain in the fiber, is manifest as a change in frequency of the scattered light. (These parameters change the acoustic velocity, and hence the Brillouin "Doppler" shift). By optical heterodyning, the frequency shifts can be down-converted to a more convenient electronic frequency, and are then in a particularly convenient form for accurate logging. Unfortunately, the Brillouin method has yet to be developed into a practical sensor. Apart from the relatively recent conception of the method, a further significant factor is the additional complexity of the optical system when compared either to the conventional OTDR or its Raman version. In addition, the limited distance resolution of the Brillouin system presents a significant drawback. The Brillouin linewidth effectively restricts the distance interval to several tens of metres, and will remain a problem unless satisfactory solutions are found.

3.5.3 Time Domain Fluorescence Monitoring

The re-emission reduce space spectrum of many fluorescent materials exhibits a significant temperature variation. It has been proposed that, in order to construct a sensor, an optical arrangement similar to that used for Raman OTDR could be constructed, using a laser source as before, but with the detector filters now selected to examine regions of the fluorescent decay spectrum of the fiber (reference 30, figure 21). In order to produce a distributed temperature sensor, wavelengths which exhibit the maximum possible temperature variation should be used. The potential attraction of the method,first proposed by Dakin (1984), is that the fluorescent quantum efficiency may be many orders of magnitude higher than that for Raman scattering, and higher doping levels can be used in order to greatly enhance the signals in short distributed sensor systems. However, there remains a problem with the availability of suitable fibers.

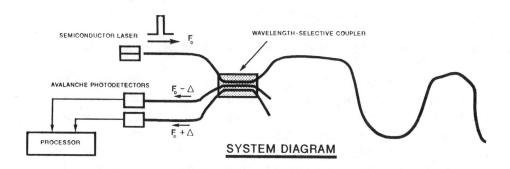

Figure 21   System diagram for distributed sensing using fluorescent OTDR

Silica-based optical fibers, having appropriate doping for high fluorescent efficiency, have been prepared using rare-earth doping (reference 31). Unfortunately, these are likely to have limited distance resolution in a fluorescent OTDR system, due to the long fluorescent lifetimes. Reduction in the lifetimes of the excited states of fluorescent dopants in glasses may generally be achieved by increased coupling to non-radiative processes, but this also reduces the fluorescent efficiency. Polymer fibers are believed, by the present author (reference 32), to offer more promise in short distance systems, as these may be doped with organic dye materials having an excellent combination of high quantum efficiency ($\approx 50\%$ or better) and short fluorescent lifetimes (of the order of only a few nanoseconds). Of course, polymer fibers will only have a limited temperature range, a feature which could be a major problem for many applications.

A theoretical paper (reference 33) has made a comparison of the performance expected from distributed temperature sensors based on the various techniques of temperature-dependent absorption, scattering, Raman scattering and fluorescence. Although fluorescent doping necessarily increases the loss in optical fibers, it was found that this would, for short distance operation, be more than compensated by the much higher light levels expected with fluorescence. If suitable glass fibers can eventually be produced it is likely to be an attractive future method for distributed thermometry.

3.6 Backscattering Systems using other Modulation Methods

The main methods discussed above have used a pulsed source, an approach used in first-generation electronic radar systems. (reference 34). This approach has the disadvantage that a low duty cycle signal is transmitted, leading to limitations in the mean power level. We shall now discuss three means by which the duty cycle of the transmitted signal can be increased. Each of these methods has been applied in sophisticated modern radar systems before finding use in experimental optical sensors.

3.6.1   Methods using a Pseudo-Random Encoded Source Modulation

The first method is that of encoded amplitude modulation of the source, using an orthogonal binary code sequence (reference 35). There are numerous binary codes that can be used for this modulation. The basic requirement is that the code should have a poor correlation with itself (ie. an autocorrelation function close to zero) for all conditions, except for the one where precisely-time-synchronised sequences are compared. Under this latter condition, the autocorrelation function has a maximum value. The effect of using such a code is to increase the duty cycle of the source to 50%, compared to typical values of 0.1% or less for the pulsed system.

In order to decode the backscatter signal with such a code, it is necessary to correlate the returning signal with a sample of the transmitted code sequence. The signature can be recovered by sweeping the relative delay between the reference code and the detected signal. Alternatively, a parallel processor system can obtain correlation signals in a series of multiplier cells, each cell corresponding to a measurement range determined by the delay of the reference signal applied to it.

Provided the averaging systems are of similar type, the coded system gives a signal/noise ratio improvement of the order of $\sqrt{N/2}$, where N is the number of discrete, resolvable, range cells. With finite length codes and imperfect waveforms, care must betaken not to introduce additional artifacts in the correlation signal which fail to correspond to real features in the backscattered signal. It has been found that the use of complimentary code sequences can lead to significant improvements in signal distortion in encoded OTDR systems (reference 36).

It should be emphasised that the encoded signal approach is applicable both to conventional OTDR (reference 35) and to Raman OTDR (reference 37).

3.6.2 Optical Frequency Domain Reflectometry (OFDR)

A second method of increasing the duty cycle is to "chirp" the optical source. This involves applying a periodic frequency modulation to the source, which should be a device, such as a laser, having a reasonably narrow instantaneous bandwidth. The preferred frequency-modulating waveform has a "sawtooth" variation with time, with a fast 'flyback'. This method is essentially similar to the frequency-modulated carrier wave (FMCW) technique used in radar systems. If an OFDR system is operated in backscattering mode, in a continuous monomode fiber, the beat signal produced at the detector increases in frequency, in direct proportion to the distance from which the light is retroscattered (references 38, 39). If the detected beat signal is examined with a conventional electronic spectrum analyser, the power in each frequency interval represents the level of scattered light received from a short section of fiber, situated at a distance corresponding to the frequency offset observed. The minimum theoretical range resolution, ΔR, possible, assuming a perfect linearly-chirped source (and a sufficient signal to noise ratio) is given by Kingsley and Davis (reference 39) as:-

$$\Delta R = 2V_g / \Delta f$$

where V_g is the velocity of light in the fiber, and Δf is the peak-peak frequency deviation of the optical source.

As the frequency-slew rate of current-ramp-driven semiconductor laser diodes may be very high (100 GHz/s is easily achievable), and the frequency resolution of commercial electronic spectrum analysers is a few Hz or less, the technique has a far superior distance resolution capability than OTDR methods. The above equation predicts a typical distance resolution of the order of 1mm, taking into account the reduced velocity of light in the fiber. Kingsley and Davies have suggested using the technique to perform distributed measurements in integrated optical wave-guide circuitry (reference 40).

Unfortunately, a major potential problem with OFDR is caused by uncertainties in the coherence function of the source. This coherence function modulates the received spectrum and therefore distorts any spatial variation of scattering that it is desired to observe. Some recent research directed towards converting this problem into a virtue and actually use the changes in the coherence to determine changes in the scatter function (reference 41).

Although several publications have reported OFDR results for fiber attenuation monitoring, none have so far used the method for distributed sensing of parameters external to the fiber.

3.6.3 Sub-carrier Frequency Domain Reflectometry (SCFDR)

The third method of encoding is to amplitude modulate the source with a chirped electronic signal (reference 42). This is essentially the same as the OFDR method, except that the sub-carrier is modulated rather than the source. This gives considerable more control of the source characteristics, as chirping of an electronic oscillator with a sawtooth modulation function is somewhat easier to achieve in practice. In addition, there is no longer a requirement for the source to have a long coherence length, as, unlike the OFDR method, no optical interference is necessary. Of course, the maximum frequency sweep and frequency-slew rate possible with the electronic sub-carrier method is generally less than that with an optical source, so the range resolution will generally be less. Despite several interesting research papers, the OFDR and SCFDR methods have, unlike the OTDR, not found significant application in commercial instruments.

4. TRANSMISSIVE DISTRIBUTED SENSING SYSTEMS

We shall now consider sensing systems where only transmitted light is monitored. The advantage is that the signal strengths are much greater than those for weak backscattered signals. The disadvantage is that there is a much smaller difference in propagation delay between modes of propagation. As mentioned in the section 1 introduction, it is necessary to have the signals carried in two modes, so that propagation delay differences can be derived, and the location of measurand-induced changes can be computed. We shall now consider various methods which have been reported. All are still at the research stage, none having yet been taken to the stage of commercial instrumentation.

4.1 The Transmissive Frequency-Modulated Carrier Wave (FMCW) Method for Disturbance Location

The optical FMCW method may be used to locate points where mode coupling in a fiber has occurred, provided that the fiber is capable of supporting two modes having significantly different phase velocities. The frequency of the source is chirped, as with the OFDR method, and the chirped signal is fed into one mode of a two-moded fiber. Any external disturbance, capable of cross-coupling energy from the initially-excited single mode, will produce, on a detector situated at the far end of the fiber, a heterodyne or beat signal. This beat signal is at a frequency equal to the difference between the optical frequencies of the signals incident on the detector. Because of the frequency-ramped nature of the source, the difference frequency detected is proportional to the distance from the source, at which mode coupling to the second mode has taken place. This approach, first suggested by Franks et al. (reference 43), is depicted in figure 22. Their particular implementation used a birefringent fiber, with the transmitted signal being launched into only one of the two principal polarization modes. The disturbance to be monitored was a transverse pressure applied to the fiber, which caused coupling of part of the propagating energy into the orthogonal polarisation mode.

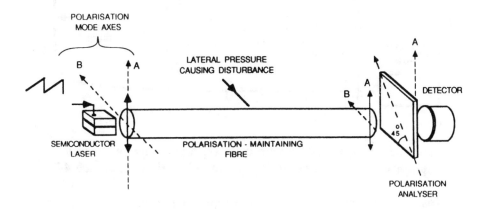

**Figure 22 Transmission FMCW distrubance location
sensor (reference 43)**

A convenient attribute of the technique is that the relatively close velocity matching between the polarization modes, even in so called high birefringence fiber, allows the FMCW technique to be used over lengths far in excess of the coherence length of the source. Two potential difficulties exist with the approach, however. Firstly, mechanical strains of certain critical magnitudes may cause coupling of power from one polarization mode to the other, and then completely back again, resulting in no net energy transfer, and hence no beat signal. Secondly, disturbances causing strain in a direction aligned exactly with either of the polarisation axes of the fiber will cause no mode coupling. Otherwise, except for these somewhat unlikely conditions, the technique appears to a simple and elegant method of locating the position of disturbances on a continuous fiber. Attractive applications include the detection and location of sources of noise or vibration, for example for fracture location in materials or for intruder location (figure 7).

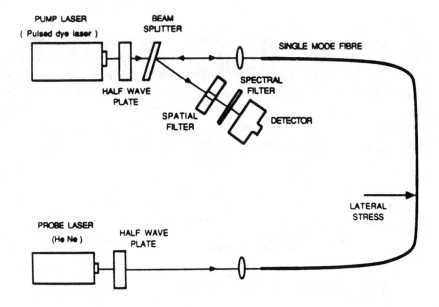

Figure 23   Sensor system with counterpropagating probe and pump-pulse signals, based on Raman amplification (reference 44)

4.2   Distributed Sensing Using an Amplifying Fiber, with Counter propagating Probe and Pump-Pulse signals

If an optical signal from a steady CW source is transmitted through a fiber to a detection system, the power level received will be dependent on the total attenuation in the fiber. If the fiber is capable of amplification, the results can be more interesting. For example, if an intense optical pulse is transmitted in the optical fiber, in the opposite direction to the CW signal, the detected signal will be affected by any nonlinear gain processes which may be created by the effects of the pump.

The first example of such a system was reported by Farries and Rogers (reference 44). This monomode fiber system used a reverse-travelling pulse from a Nd:YAG-pumped dye laser at 617 nm to provide Raman gain in a fiber. The gain was monitored using a continuous 633 nm forward-travelling signal from a helium-neon laser source. The Raman gain is strongly sensitive to the relative polarization of pump and probe signals. The arrangement of figure 23, in a fiber of low intrinsic birefringence, is capable of detection and location of lateral stresses, as these cause polarisation mode conversion, which modifies the Raman gain.

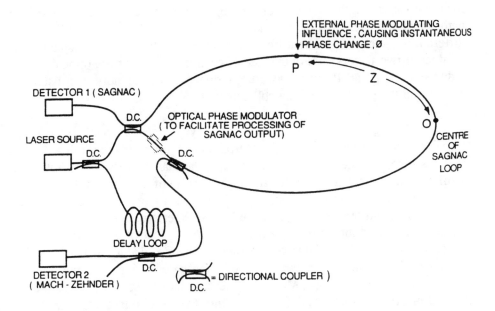

**Figure 24   Modified Sagnac interferometer with Mach-Zehnder
reference interferometer for disturbance location
(reference 46)**

The technique has considerable academic interest, as it was the first of a new class of sensors. However, it suffered from a number of practical disadvantages because it uses inconvenient optical sources, and is likely to be critically dependent on the pump power level of the dye laser source and to suffer from undesired polarization drift under the influence of normal environmental conditions on other sections of the fiber.

4.3 Disturbance Location using Sagnac Interferometers

The fiber optic Sagnac interferometer consists, in its simplest (See footnote) form, of a monomode fiber loop and a directional coupler. The arrangement allows the launching of counter-propagating beams into the loop, from a single source, and the detection of the superimposed waves returning to a detector via an exit port of the same coupler (figure 24). Such arrangements have been studied extensively for use in optical fiber gyroscope systems, although additional components are usually inserted to produce a reciprocal configuration.

A major source of phase error in optical gyroscopes (reference 45) may occur if rapid changes in optical path length, due to thermal or mechanical effects on the fibers, are allowed to occur at the position, P, away from the geometrical center, O, of the fiber length used to produce the coil. The reason for the error is that the two counter-propagating beams encounter the varying path length changes at different moments in time, and therefore will suffer different phase changes. The difference in phase change is proportional to the product of two factors; firstly the rate of change, $d\phi/dt$ of the optical signal induced at the point, P, by the external influence and, secondly, the distance, z, between the point, P, and the coil center, O.[†]

Although the effect is a drawback for gyroscopes, the method has been researched as a means of locating disturbance in a fiber (reference 46). In the first instance, in order to prove the concept, thermal changes were chosen as a suitable time-varying influence, although location of mechanical disturbance will probably have more potential applications. In order to calculate the distance, z, from the imbalance in the Sagnac interferometer (which, as has already been described, is proportional to the product of z and $d\phi/dt$), the method requires a knowledge of the rate of change, $d\phi/dt$, induced by the influence. This quantity cannot be readily measured using only the Sagnac loop. However, an additional Mach-Zehnder interferometer may be introduced, by adding an additional fixed-delay fiber link from the source and combining the output of this with a signal extracted from one of the counter-propagating beams in the Sagnac loop (figure 24). The output from the Mach-Zehnder interferometer gives an output proportional to change, ϕ, and differentiation of this phase output (conveniently performed in these measurements by a frequency-counting system, which monitors the number of interference fringes passed through per second at the detector) yields the required rate of change, $d\phi/dt$. Simple division of the Sagnac phase offset by $d\phi/dt$ finally gives the desired distance, Z, of the point of disturbance, P, from the center point, O. Experimental results for location of the thermal disturbance are shown in figure 25.

With a suitable nonreciprocal Sagnac configuration, and accurate interferometric processing, the technique shows promise for accurate location of quite modest disturbance levels. The location capability of such a system has been analysed in reference 47. More recent theoretical proposals have been made by Udd, to improve the configuration of the Sagnac disturbance-sensing system (reference 48). These modifications enable are intended to allow the sensor to be used with broader band sources and reduce the effects of back-reflected light to the source.

[†] For accurate low-drift operation of gyroscopes, reciprocity of counter-propagating beams is necessary. To ensure this, an addition single polarization mode filter is necessary, along with additional coupler to permit its inclusion into the system (reference). The sensor experiments described here would also benefit from such an arrangement, although the preliminary reported results were, for convenience, carried out with the simpler Sagnac loop.

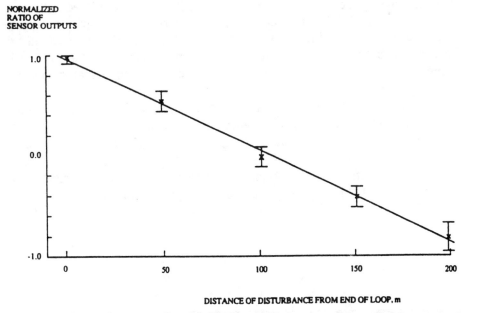

NORMALIZED
RATIO OF
SENSOR OUTPUTS

DISTANCE OF DISTURBANCE FROM END OF LOOP, m

**Figure 25   Response of Sagnac location system to thermal
disturbances located at different points in the Sagnac loop**

4.4 Transmissive Location Systems using White Light Interferometry

A recent method of location of disturbance in a two-moded fiber is based on white light interferometry (reference 49). If a broadband source, such as an LED, is directed along two independent monomode paths and then recombined, interference fringes can only be observed when the paths are almost equal. For path differences much greater than the coherence length of the source, the superposition of multiple, differently-phased, fringes, from each individual frequency component of the broadband source, will lead to a total loss of visible interference. For a typical LED, the coherence length is of the order of only 30 μm, so quite small intervals in optical path length can be resolved.

A two-moded fiber having significantly different mode velocities does not have to be very long in order to generate path differences which are sufficient to prevent visible interference fringes from a broadband light source, when the mode outputs are combined at the end of the fiber. For a high-birefringence fiber with 2 mm beat length, a length greater than 100 mm will suffice to lose coherence. However, if a separate twin-path interferometer is placed at the exit end, and the optical path length differences is matched to that in the fiber, then two equal length interfering paths to the detector can be generated. When the external interferometer brings the system close to equal optical path length conditions, intensity fluctuations can be observed as the path is changed (figure 26; reference 49). If the external interferometer is a free space type, such as a Michelson arrangement, quite small differences in this external path can balance much longer lengths of fiber, as the latter will generally have only a small difference propagation velocity between modes.

To form a disturbance location system, broadband light is launched into only one mode of a high-birefringence fiber. Pressure along the length can couple part of the propagating energy into the other mode. From an observation of the time at which visible fringes are observed, a scanned-path-difference external interferometer can now determine the distance between the in-fiber coupling point and the output end of the fiber.

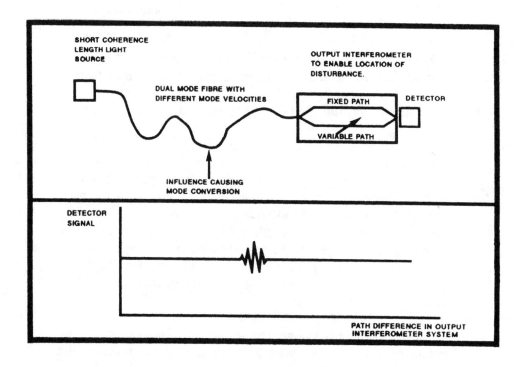

Figure 26 Illustration of white light interferometry for location of disturbance in dual-mode fibres (Kotrotsios and Parriaux 1989)

CONCLUSIONS

The current status of distributed fiber sensors has been reviewed. The field is still one of rapid development, but commercial versions of the distributed temperature sensor, of the type first reported in reference 24, are now becoming available from several industrial sources. It is likely that practical systems will eventually be developed to cover other important application areas, such as the distributed measurement of mechanical strain, chemical concentration, electrical and magnetic fields. There is still a considerable challenge in the technology and many areas where new innovations are required.

REFERENCES

1. "Fiber waveguides: a novel technique for investigating attenuation characteristics" Barnoski, M.K., Jensen S.N., Applied Optics 15 (1976) pp 2112-2115.

2. "Instrumentation principles for OTDR", Healey P., J. Phys. E (Sci. Instr.), 19 (1986) pp 334-341.

3. "Microending losses of single-mode, step-index and multimode parabolic-index fibers", Marcuse E., Bell systems Tech. Jour 55 (1976) pp 937-955.

4. Data sheet for "Hergolite" pressure sensitive cable, Herga Ltd, Bury St Edmunds, UK.

5. "Fibre optic strain measurement for structural integrity monitoring". Brunzma A J A, Van Zuylen. P., Lamberts C.W., de Krijger A.J.T., Proc OFS 84 Int Conf, Stuttgart (publishers: Berlin VDE) pp 399-402.

6. "Application of optical fibre waveguides in radiation dosimetry", Gaebler W. and Braunig D., Proc 1st Int Conf on Optical Fibre Sensors, London (London: IEE) (1983) pp 185-9.

7. "Fibre optic temperature distribution sensor", Hartog A.H. and Payne D.N., Proc IEE Colloq Optical Fibre Sensors (London: IEE) (1982).

8. "Differential absorption distributed thermometer", Theocharous E., Proc 1st Int Conf on Optical Fibre Sensors, London (1983) pp 10-12.

9. "Optical fibre sources, amplifiers and special fibres for application in multiplexed and distributed sensor systems", Cowle G.J., Dakin J.P., Morkel P.R., Newson T.P., Pannell C.N., Payne D.N. and Townsend J.E., Proc O/E Fibers 91, Boston USA (Sept 1991) (SPIE vol. 1586) pp 130-145.

10. "Optical fibres for cryogenic leak detection", Pinchbeck D., Proc Conf on Electronics in Oil and Gas, London (London: Cahners Exhibits Ltd) (1985).

11. "A plastic-clad silica fiber chemical sensor for ammonia", Blyler Jnr. L.L., Ferrara J.A., MacChesney J.P., Proc. OFS Int Conf., New Orleans, USA, 1988 (pub O.S.A.) pp 369-372.

12. "Distributed liquid sensor using eccentrically clad fiber", Yoshikaw, H., Watanabe M. and Ohno Y., Proc OFS '88, Tokyo, 1988.

13. "Polarisation optical time domain reflectometry", Rogers A., Electron. Lett. 16 (1980) pp 489-90.

14. " POTDR : Experimental results and application to loss and birefringement measurements in single mode fibres", Hartog A.H., Payne D.N. and Conduit A J., Proc 6th ECOC, York (Published London: IEE) (1980) (post deadline paper).

15. "Backscattering measurements of bending-induced birefringence in single mode fibres", Kim B.Y. and Choi S.S., Electron. Lett. 17 (1981) pp 193-195.

16. "Measurement of magnetic field by POTDR", Ross J.N., (1981), Electron. Lett. 17 pp 596-597.

17. "POTDR a technique for the measurement of field distributions", Rogers A.J., Appl. Optics 20 (1981) pp 1060-1074.

18. "Fading in heterodyne OTDR", Healey P., Electron. Lett. 20 (1984) pp 30-32.

19. "Distributed optical fibre sensors for the measurment of pressure, strain and temperature", Rogers A.J., Physics Reports 169 (1988) pp 99-143.

20. "152 Photons per bit detection at 2.5 G Bit/s using an Erbium fibre preamplifier", Smyth P.P., Wyatt R., Fidler A., Eardley P., Sayles A., Blyth K., Craig-Ryan S., Proc ECOC '90, Amsterdam, Sept (1990) pp 91-94.

21. "Noise in Erbium-doped amplifiers", Laming R.I., Morkel P.R., Payne D.N. and Reekie L., Proc. ECOC 88, Brighton, UK, Published by IEE, London (1988), pp 54-57.

22. York Sensors (Chandlers Ford, UK). Data sheet for DTS 80 commercial distributed temperature sensor.

23. "Reference for Raman spectrum measured by CERL Research Labs, Leatherhead, U.K., in joint publication with CERL (reference 24).

24. "Distributed anti-Stokes ratio thermometry", Dakin J P., Pratt D.J., Bibby G.W., Ross J.N., Proc 3rd Int Conf on Optical Fiber Sensors, San Diego (1985) (post deadline paper).

25. "Distributed optical fibre Raman temperature sensor using a semiconductor light source and detector", Dakin J.P., Pratt D.J., Bibby G.W. and Ross J.N., Electron. Letts 21 (1985) pp 569-70.

26. "Distributed temperature sensing in solid-core fibres", Hartog A.H., Leach A.P. and Gold M.P., Electron Letts 21 (1985) pp 1061-1062.

27. York Sensors DTS II (earlier commercial instrument).

28. "Exploitation of stimulated Brillouin scattering as a sensing mechanism for distributed temperature sensors and as a means of realising a tunable microwave generator", Culverhouse D., Farahi F., Pannell C.N. and Jackson D.A., Proc OFS '89, Paris 1989, pp 552-559 (Springer Verlag, 1989 ISBN 3-540-51719-7).

29. "Optical fibre sources, amplifiers and special fibres for application in multiplexed and distributed sensor systems", Cowle G.J., Dakin J.P., Morkel P.R., Newson T.P., Pannell C.N., Payne D.N. and Townsend J.E., Proc O/E Fibers 91, Boston USA (Sept 1991) (SPIE vol. 1586) pp 130-145.

30. UK Patent Application GB 2156513A, Dakin J.P., (priority date 28 March 1984, published 9 October 1985).

31. "Fabrication of low-loss optical fibres containing rare-earth ions", Poole S.B., Payne D.N. and Fermann M.E., Electron. Lett. 21 (1985) pp 737-8.

32. "Multiplexed and distributed optical fibre sensor systems", Dakin J.P., J. Phys. E, Vol 20 (1987) pp 954-967.

33. "Fibre-optic distributed temperature measurement - a comparative study of techniques", Dakin J.P. and Pratt D.J., Proc IEE Colloq. on Distributed Optical Fibre Sensors London 1986 : IEE Digest No 1986/74, pp 10/1-10/4.

34. "1970 Radar Handbook" Skolnik M I. New York: McGraw-Hill (1970).

35. "An optical time domain reflectometer with low power InGaAsP diode lasers", Subdo A.S., IEEE J. Lightwave Tech., LT-1 (1983) pp 616-618.

36. "A new technique in optical time domain reflectrometry", Newton S.A., Opto Elektronic Magazin 4 (1988) pp 21-33.

37. "Novel signal processing techniques for enhanced optical time domain reflectometry sensors", Everard J.K.A., Proc SPIE vol 798 (1987) pp 42-46.

38. "Optical FM applied to coherent interferometric sensors", Uttam D. and Culshaw B., Proc IEEE Colloq on Optical Fibre Sensors (1982) (London: IEE), IEE Digest No 1982/60.

39. "Multiplexed optical fibre interferometers: an analysis based on radar systems", Al Chalabi S.A., Culshaw B., Davies D.E.N., Giles I.P. and Uttam D., Proc IEE 132 (1985) pp 150-6.

40. "OFDR diagnostics for fibre and integrated-optic systems", Kingsley S.A. and Davies D.E.N., Electron. Lett 21 (1985) pp 434-5.

41. "Optical coherence domain reflectometry by synthesis of coherence function", Hotate K., Kamatani O., Proc OE/Fibers '91, Distributed and Multiplexed Fiber Optic Sensors, Boston, USA 1991 : Proc SPIE 1586 pp 32-45.

42. "Fault location in optical fibres using optical-frequency-domain reflectometry", Ghafoori-Shiraz H. and Okoshi T., J Lightwave Tech LT-4 (1986) pp 316-22.

43. "Birefringent stress location sensor", Franks R.B., Torruellas W., Youngquist R.C., SPIE Vol 586 (1986).

44. "Distributed sensing using stimulated Raman interaction in a monomode optical fibre", Farries M.C. and Rogers A.J., Proc 2nd Int Conf Optical Fibre Sensors OFS '84 Stuttgart (Berlin: VDE) (1984) pp 121-32.

45. "Thermally-induced nonreciprocity in the fiber-optic interferometer", Shupe D.M., Appl. Optics <u>19</u> (1980) pp 654-655.

46. "A novel distributed optical fibre sensing system, enabling location of disturbances in a Sagnac loop interferometer", Dakin J.P., Pearce D.A., Wade C.A., Strong A., Proc OE Fiber 1987 San Diego (1987). Proc SPIE Vol <u>838</u>, paper 18.

47. "A novel distributed optical fibre sensing system, enabling location of disturbances in a Sagnac loop interferometer", Dakin J.P., Pearce D.A., Strong A., Wade C.A., Proc ECOC/LAN Int Conf, Amsterdam, June 1988.

48. "Sagnac distributed sensor concepts", Udd E., Proc OE Fibers 1991, Distributed & Multiplexed Fiber Optic Sensors, Boston, USA (1991) SPIE Vol <u>1586</u> pp 46-52.

49. "White light interferometry for distributed sensing on dual mode fibers", Katrotsios G. and Parriaux O., Proc OFS '89, Paris 1989 : Pub. Springer Verlag ISBN 3-540-51719-7 (1989) pp 568-574.

MULTIPLEXED FIBER OPTIC SENSORS

Alan D. Kersey

Optical Techniques Branch, Code 5670
Naval Research Laboratory
Washington, D.C. 20375

ABSTRACT

A wide range of multiplexing techniques for fiber optic sensors have been proposed and demonstrated over the past 10 years. In many cases, systems utilizing multiplexed sensors have under gone field trails which have successfully proven the technology. This paper reviews this technology, and discusses recent research efforts in the area.

1. INTRODUCTION

The ability to multiplex sensors is an important issue in many of the proposed application areas for fiber optic sensors. Whether the application involves high sensitivity military sensor systems, industrial process control sensors, chemical sensing, or environmental and structural sensing, the use of multiplexing techniques can be beneficial in regard of a number of system aspects including reduced component costs, lower fiber count in telemetry cables, ease of E/O interfacing, and overall system immunity to EMI. The development of efficient multiplexing techniques can thus be expected to lead to general improvements in the competitiveness of fiber sensors compared with conventional technologies in a broad range of application areas.

This paper reviews the development of multiplexing techniques for fiber sensors, including simple serial arrays of sensors based on optical time domain reflectometry (OTDR) processing concepts, to highly sophisticated interferometric fiber sensors. Recent developments in the area are also discussed.

2. SERIAL POINT SENSOR (QUASI-DISTRIBUTED) NETWORKS

The simplest form of multiplexed sensor system involves the serial concatenation of point or 'quasi-point' fiber sensors in a linear array. This type of system can be interrogated using OTDR signal processing [1], and is an extension of fully distributed fiber sensing techniques (Dakin [2], these proceedings) to the interrogation of a finite number of discrete sensors. Figure 1 shows such an implementation of a quasi-DFOS (QDFOS) system. Various sensing methods and addressing techniques have been used to implement quasi-distributed sensor systems. For example, modified fiber sections with sensitized optical properties

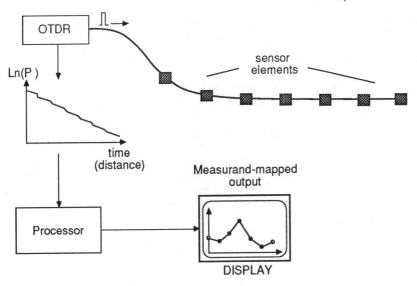

Figure 1. Quasi-distributed serial array using OTDR processing

can be spliced into a long fiber at certain intervals to provide localized variations in the loss, backscatter intensity, polarization, fluorescence intensity, etc. This is different from intrinsic-distributed fiber sensing in that the measurand can be determined at a finite number of locations only, and not continuously along the fiber path. Alternatively, discrete non-fiber sensor elements which vary in transmittance or reflectance with the measurand field can be incorporated into the fiber line. Such an arrangement for distributed temperature sensing was demonstrated at a early stage in the development of QDFOS technology [3]. This system used ruby glass sensor elements, the attenuation of which increase with temperature for light of wavelength ~ 600 to 620 nm (absorption edge shift rate ~ 1.2 Å/°C). OTDR type interrogation of the system was used to determine the loss at each sensor element, and a second wavelength removed from the absorption edge was used to provide a temperature independent reference output. Other materials, such as semiconductors are also suitable for this approach, as are fibers doped with certain elements, e.g. Hollium, Neodymium. The major limitations of this system, and similar approaches [4,5] is the fact that the attenuation is accumulative; the light levels at the most distal sensor thus depends on the measurand at each sensor along the fiber. This places demanding requirements on the dynamic range of the detection system and limits the number of sensors which could be used in a practical system. This is also true, but to a lesser extent, for systems based on reflective sensing elements [6].

3. FIBER BRAGG-GRATING BASED SENSORS

Intra-core fiber Bragg grating (FBG) sensors have attracted considerable interest over the past few years because of their intrinsic nature and wavelength-encoded operation. The gratings are holographically written into Ge-doped fiber by side-exposure to UV interference patterns [7-9]. Other means for producing such gratings also exist, and other fiber dopants may be used to improve efficiency, or

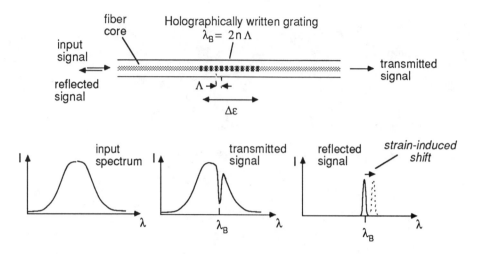

Figure 2. Fiber Bragg grating sensor system

alter the required writing wavelength. These sensors will prove to be useful in a variety of applications, in particular, in the area of advanced composite materials, or 'smart structures' where fibers can be embedded into the materials to allow real time evaluation of load, strain, temperature, vibration etc. Figure 2 shows the generic sensing concept involved for a single sensor element. The fiber Bragg grating (FBG) sensor is illuminated using a broadband source (BBS), such as an edge-emitting LED, superluminescent diode, or superfluorescent fiber source. The narrow wavelength component reflected by the sensor is determined by the Bragg wavelength;

$$\lambda_B = 2n\Lambda, \tag{1}$$

where n is the effective index of the core, and Λ is the period in the index modulation of the core induced by the UV exposure. Measurand-induced perturbation of the grating sensor changes the wavelength returned, which can be detected and related to the measurand field (e.g. strain) at the sensor position. The wavelength-encoded nature of the output has a number of distinct advantages over other direct intensity based sensing schemes, most importantly, the self-referencing nature of the output; the sensed information is encoded directly into wavelength, which is an absolute parameter and does not depend on the total light levels, losses in the connecting fibers and couplers or source power. The reported dependence of the (normalized) shift in Bragg wavelength with fiber strain, ε, is $(1/\lambda_B)(d\lambda_B/d\varepsilon) \approx 0.74 \times 10^{-6}/\mu$strain, where 1 μstrain is a strain of 1 part in 10^6, and a temperature dependence $(1/\lambda_B)(d\lambda_B/dT)$ of $\approx 8.9 \times 10^{-6}/°C$.

These FBG elements are ideal for multiplexed networks, and a variety of configurations have been proposed [9,10]. Figure 3 shows a generalized concept for multiplexing based on wavelength division addressing. Here, the gratings are asigned a particular wavelength range, or 'domain' for operation which do not overlap. The Bragg wavelengths of the individual grating can thus be determined

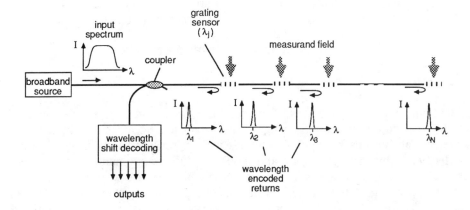

Figure 3. FBG sensor array

by illuminating the system with a broadband source and using an optical spectrum analyzer (spectrometer) to analyze the return signal. This simplest approach is practical for only a limited number of devices, simply due to the fact that the bandwidth of sources are limited, and can thus only accommodate a specific number of grating operational wavelength bands.

A means to overcome this limitation is to adopt some form of time division multiplexing (TDM) in conjunction with the inherent wavelength division multiplexing (WDM) capability of the grating sensors. Figure 4 shows a

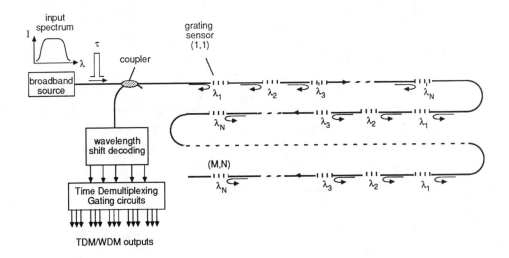

Figure 4. FBG array based on time and wavelength division addressing.

proposed concept using both TDM and WDM for addressing a large number of elements [10,11]. This type of signal processing may allow a large number of grating sensors to be interrogated in a serial array which would be of interest in applications such as embedded sensor systems for smart structures.

4. INTENSITY-SENSOR BASED NETWORKS

4.1 General

The term 'intensity based sensor' is used to describe a generic class of sensors which depend on monitoring changes some characteristic related to the detected intensity at the sensor output. Examples include sensors based on attenuation, reflectance, fluorescence signal, and modal modulation. A number of different types of branching networks have been investigated for use with intensity-based sensors, particularly those based on simple concepts such as attenuation. Sensors can be addressed using schemes based on optical analogs of conventional electronic time- and frequency- division multiplexing (TDM and FDM respectively) techniques, or by using schemes devised for use in optical communications systems such as wavelength-division multiplexing (WDM).

4.2 Time-division multiplexing

The first passive discrete-sensor network was proposed by Nelson et. al. [12] and used TDM to address a number of reflective intensity-based sensors. These sensors were spaced at different distances from the source and detector, such that a single pulse, of appropriate duration at the input to the network produced a series of distinct pulses at the output. These pulses represent time samples of the sensor outputs interleaved in time sequence, as shown in Figure 5. The required duration of the input pulse is determined by the effective optical delay of the fiber connecting the sensor elements, and repetitive pulsing of the system allows each sensor to be addressed by simple time-selective gating of the detector

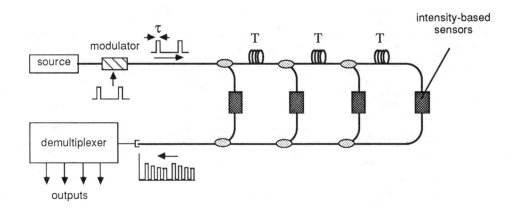

Figure 5. Time division multiplexed intensity sensor array

Network configurations for both transmissive and reflective intensity based sensors were described. Spillman and Lord have reported a self-referencing TDM intensity sensor network based on recirculating fiber loops [13]. This work has also been extended to use frequency division addressing [14].

4.3 Frequency-division multiplexing

A number of novel concepts for frequency-domain-based multiplexing schemes for intensity sensors have also been reported. Mlodzianowski et.al. [15] described a scheme in which the individual sensor information is carried not by separate beat or carrier frequencies, but by the phase and amplitude of an RF sub-carrier amplitude modulation of source light returned from a number of sensor elements. Interrogation of the system at a number of discrete modulation frequencies allows the status of each sensor to be interpreted. A system comprising three sensors has been demonstrated using this technique, and showed particularly good crosstalk performance (~ -40 dB). In another approach, the radar-based FMCW technique has also been used to allow frequency division addressing with a network of intensity based sensors [16]. In this case a chirped RF intensity modulated source is used to interrogate a number of simple reflective intensity sensors, and the detector output is electrically mixed with a 'reference' chirp signal. This produces a beat frequency associated with each sensor element, allowing frequency demultiplexing of the outputs.

4.4 Wavelength-division multiplexing

Demonstrations of wavelength division multiplexing in fiber communications systems have been reported for many years [17-19]. The use of this technique in sensor application has not, however, received much practical attention. Figure 6 shows the type of arrangement possible using WDM. The scheme, which is

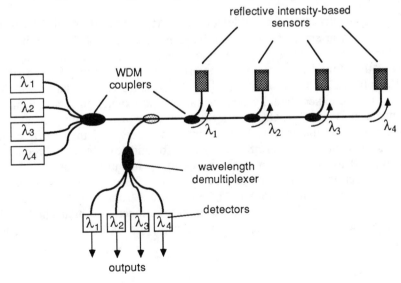

Figure 6. Wavelength division multiplexed sensor array

essentially applicable to both intensity and interferometric sensor types, is theoretically the most efficient technique possible, as all the light from a particular source could in principle be directed to a particular sensor element and then onto a corresponding photodetector with minimal excess loss. The reason for the lack of practical demonstrations of this technique is due to the limited availability of wavelength-selective couplers (splitters and recombiners) which are required to implement the technique. This combined with the complexity of the WDM fiber components (e.g. N x N star and 1 xN tree couplers) needed to build systems based on a number of sensors and the limited wavelength-selectivity of such devices are the major drawbacks of the approach. Consequently, apart from the obvious use of WDM techniques in FBG systems (Section 3), wavelength division multiplexing of large numbers of discrete intensity-based (or interfero-metric) sensors utilizing common servicing fibers may not prove to be viable, in terms of both cost and performance.

4.5 Subcarrier based multiplexing

Another technique for the multiplexing of fiber sensors which is based on subcarrier signal processing has been demonstrated. In this case, each sensor in the network is a transversal filter which consists of two fibers of unequal length connected in parallel. In response to an RF intensity-modulated source the recombined light at the output of such a filter exhibits a series of minima when the differential delay in the two fiber paths corresponds to a half-integral number of cycles of the modulation frequency. The normalized frequency response of a single sensor is given by [20,21]

$$g_i(f) = |\cos(\pi\Delta\tau_i f)|, \tag{2}$$

where f is the frequency of the modulation and $\Delta\tau_i$ is the differential delay of sensor i. For a linear array of sensors the frequency response of the combination is given by $G(f) = \Pi g_i(f)$; i.e., the product of the frequency response functions of the individual elements in the array [22]. A system of three fiber-optic differential-delay filters configured as temperature sensors based on this approach has been experimentally demonstrated. The experimental arrangement used is shown in Figure 7. Light from a SLD source, which was modulated by a voltage-controlled oscillator (VCO), was input to a series of three differential delay sensor elements. In Figure 8, curves (a) , (b) and (c) show the frequency response from 0 to 20 MHz for each sensor independently operated with the source and detection system, whereas the measured frequency response of the three-sensor network is shown in curve (d) of Figure 2 from 0 to 20 MHz and in curve (e) from 10 to 16 MHz. In order to monitor temperature-induced shifts in the null frequencies corresponding to each sensor, the null-tracking technique detailed in [23] was used.

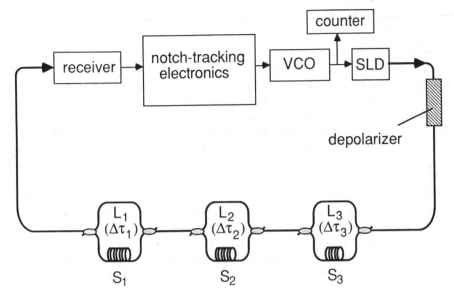

Figure 7. Subcarrier multiplexed sensor system

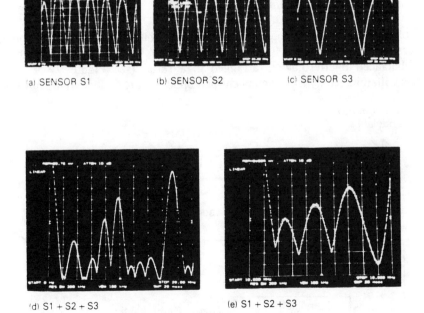

(a) SENSOR S1   (b) SENSOR S2   (c) SENSOR S3

(d) S1 + S2 + S3   (e) S1 + S2 + S3

Figure 8. Response curves for the sensor arrangement of Figure 7:
a - c individual sensors; d and e, combined transfer function.

5. INTERFEROMETRIC SENSOR MULTIPLEXING

5.1 General

Interferometric fiber sensors are being widely researched for use in a range of application areas including acoustic pressure, and magnetic and electric fields [24]. A number of different multiplexing topologies have been devised and tested by research groups working in this field. Early work during the mid to late 1980s concentrated on demonstrating the principle of operation of various multiplexing approaches such as time-division (TDM), frequency-division (FDM) and coherence multiplexing (Coh.M) using a relatively low number of sensors. In more recent years, however, arrays with up to 10 sensors [25] multiplexed on a common input/output fiber pair have been reported in the literature, representing the first demonstrations of significant multiplexing gain achieved in practical systems. Additionally, a system utilizing a hybrid TDM/WDM approach was demonstrated with 14 sensor elements supported on a single input/output fiber pair. Systems have also been taken beyond the laboratory environment: An array comprising a total of 48 networked sensors based on a FDM scheme was tested at sea in 1990.

A range of different multiplexing topologies continue to be investigated and tested by research groups working in this field. Developments in the areas of frequency, time, coherence, and code-division based systems continue to be made. The following sections discuss these developments, and recent experimental results.

5.2 Frequency division multiplexing

One of the earliest approaches developed for the multiplexing of interferometric sensors was based on the FMCW concept [26-28]. This scheme relies on the use of unbalanced interferometers arranged in a serial (see Figure 9) or parallel network illuminated by a frequency chirped optical source. Due to the inherent

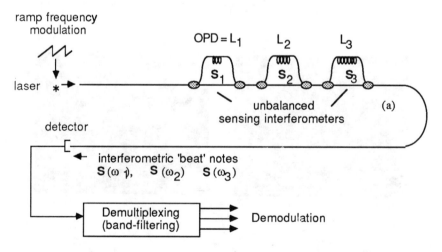

Figure 9. FMCW interferometric sensor multiplexing

sensitivity of an unbalanced sensor to input optical frequency, a beat frequency is generated at each interferometer output, the period of which depends on the frequency excursion of the chirp, the chirp rate, and the interferometer optical path difference (OPD). Assigning a different OPD to each interferometer allows the beat-frequencies associated with each sensor element to be distinct, and thus separable using band filtering. One major problem which arises with this type of multiplexing technique is cross-terms due to unwanted interferometric components arising differentially or additively between sensors, or 'ghost' interferometers arising from connecting fiber paths in conjunction with the interferometers. These cross-terms lead to sensor-to-sensor interference, or crosstalk which is a problem in most applications where the full capability of an interferometric sensor, in terms of the detection sensitivity and dynamic range, are important. These 'stray' components can be minimized using certain topologies, but cause significant design complexity for an array involving an appreciable number of sensors elements. Little experimental work has been reported on this approach since the demonstration of a three-sensor multiplexed system in 1986 [28].

A preferred FDM approach utilizes the spatial and frequency domain separation of sensor signals shown in Figures 10. In Figure 10.a., the outputs from K sensor elements all powered from a common source are 'spatially-multiplexed' onto separate fibers. In Figure 10.b. the outputs from J sensors, which are independently illuminated by separate sources, are combined onto a single output fiber. Using phase-generated-carrier (PGC) interrogation [29], with each laser

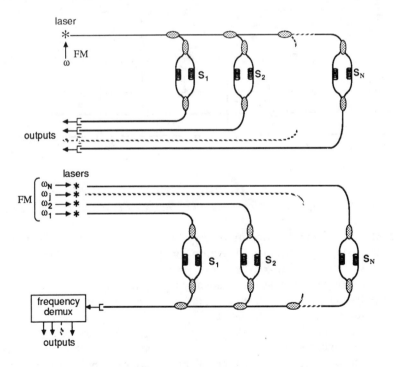

Figure 10. a) 'Spatial' and b) frequency domain addressing of interferometers

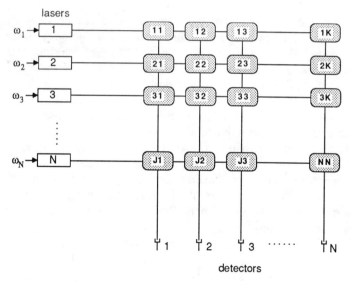

Figure 11. JxK matrix array configuration based on the spatial and frequency domain multiplexing concepts of Figure 10.

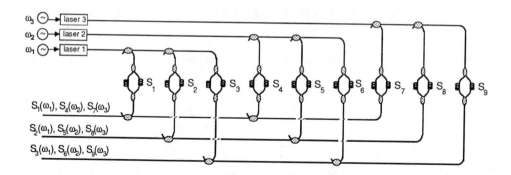

Figure 12. Practical implementation of the JxK FDM matrix array system

operated at a different 'carrier' frequency, the sensor outputs in Fig 10.b can be separated using synchronous detection or band-filtering. Combining these techniques allows a matrix-type array [30] to be configured, which contains N = JxK sensor elements, as schematically represented in Figure 11. This system is somewhat unique in that the operation of the remote PGC interrogation (demodulation) scheme automatically provides both the demodulation and demultiplexing functions, provided the sources are modulated at different carrier frequencies. Figure 12 shows a practical implementation of this type of array for a 3x3 (9 sensor) system. This type of array has been shown to be capable of providing good phase detection sensitivity and low crosstalk for systems involving up to eight sensor outputs combined onto a single output fiber.

This type of array is the most highly developed topology demonstrated to-date; an array comprising 48 acoustic sensors was successfully demonstrated in a sea test under a joint NUSC/NRL advanced technology demonstration program in 1990 [31].

5.3 Coherence multiplexing

The basic principle of the coherence multiplexing concept for interferometric sensors is shown in Figure 13. Although there was initially significant interest in coherence multiplexing [32-34], problems associated with crosstalk, excess phase noise and poor power budget have limited the practical use of this approach with interferometric sensors. Nevertheless, strong interest in the use of this approach remains for other less demanding applications; for example, for use with interferometric sensors configured to detected quasi-static (DC) measurands. A two-element multiplexed temperature sensor system based on a wavelength-modulation scheme for monitoring interferometric OPD [35] has been reported. In this case the ultra-high phase detection sensitivity normally attainable in interferometric sensor systems is not required, and crosstalk levels of ~ –40 dB maybe tolerable. The coherence-addressing and multiplexing of polarimetric sensors has also been recently reported [36,37].

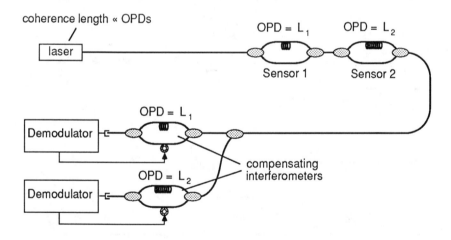

Figure 13. Coherence multiplexed interferometric array

5.4 Time division multiplexing (TDM)

Considerable interest has been directed towards the experimental demonstration and evaluation of multiplexing topologies based on time division addressing [38-45]. This work led to the development of a number of array architectures based on serial and parallel topologies, of the type shown in Figures 14 and 15 respectively. The system of Figure 14.a, referred to as a reflectometric sensor array, was the first interferometric TDM array to be demonstrated. This configuration utilized in-line partially-reflective fiber-splices, or in-fiber reflectors [38], between fiber sensing coils each of length L, to form in-line interferometric elements, which can be interrogated with pulsed operation of the source and a compensating interferometer of delay equal to the round trip delay ($T_d = 2L/nc$) between reflectors. Providing the width of the input pulse, τ, is less than T_d, an interferometric signal from each element in the array can be generated in time sequence at the compensator output. The initial demonstration of this concept utilized a differential delay heterodyne interrogation approach, where two pulses of differing optical frequencies and separated in time by T_d were launched into the array, such that at the output pulses reflected from consecutive partial reflectors in the array overlapped to produce heterodyne beat signals associated with each sensing element, without the need for a compensating interferometer. An array of six acoustic sensors based on this approach has been field tested [39].

A similar type of operation, but configured in a transmissive , was achieved using the tapped serial array (TSA) [40,41] topology shown schematically in Figure 14.b. The system is based on the use of low-coupling ratio couplers which tap off a fraction of the light in the input fiber to an output bus as shown. Fiber coils in the input fiber serve both as delay and sensing elements. Pulses obtained from the series of N+1 (for N sensors) tap points are separated in time if the delay in each sensor coil is longer than the width of the input pulse (τ) to the system. These output pulses are then coupled to a compensating interferometer which splits the

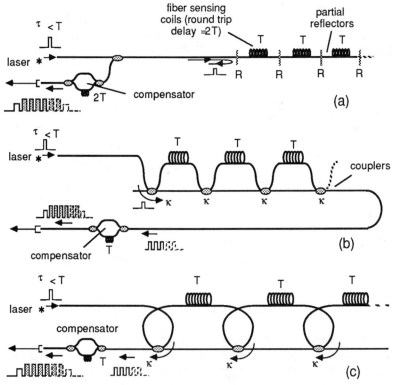

Figure 14. Time division multiplexed serial arrays

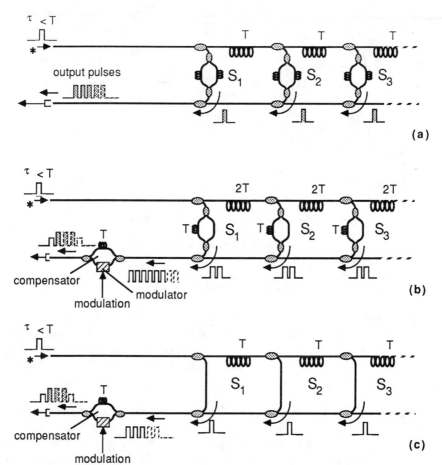

Figure 15. Time division multiplexed parallel (ladder) arrays

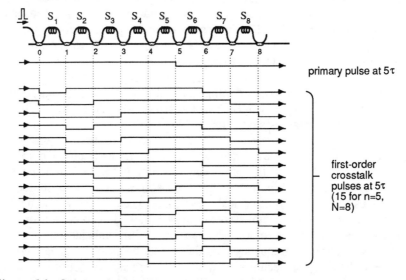

Figure 16. Origin of the multi-coupling crosstalk paths in the TSA system

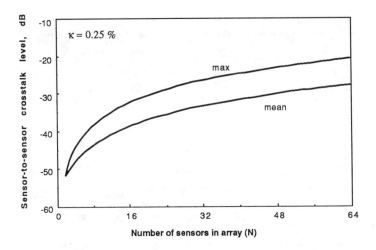

Figure 17. Crosstalk in the TSA

pulse stream into two components, delays one by a period equal to the sensor coil delay T, and subsequently recombines the signals. This forces components of adjacent pulses to overlap, resulting in interference signals which can be monitored at the detector. An array of eight sensors has been experimentally demonstrated using this approach [41].

Due to the possibility of multiple pulse reflections between partial reflectors, and coupling between the input and output fibers at the tap points, both the reflectometric and TSA array topologies give rise to multiple pulse interactions which leads to crosstalk. Figure 16 shows an example of the origin if the type of effect in the TSA system. This problem has been addressed both experimentally and theoretically at NRL. We have found excellent agreement between predicted and observed crosstalk levels between sensor elements in an eight sensor array [41,42]. As shown in Figure 17, the results of this type of analysis shows that time-average sensor-sensor crosstalk levels can be < 30 dB for an array of 25 sensors using couplers with a 0.25 % coupling ratio.

The 'recursive lattice' [43] array topology of Figure 14.c is functionally identical to the reflectometric array, giving rise to the same crosstalk effects. This array has not been experimentally tested to date.

More recent work in TDM systems has concentrated on 'ladder' array configurations [25,44,45] of the form shown in Figure 15. A ten-element array based on the topology of Figure 15.a has been successfully demonstrated. This topology does not lead to direct optical crosstalk between the sensor elements, and phase detection sensitivities comparable to those obtained in single sensor systems have been achieved. Interferometers which are slightly unbalanced to allow for passive demodulation via frequency modulated laser based phase generated carrier or synthetic heterodyne techniques are used. Phase detection sensitivities obtained with the array ranged from 12 to 18 μrad/√Hz at 1 kHz. Crosstalk

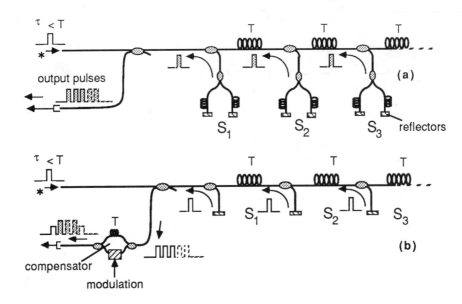

Figure 18. Time division multiplexed serial arrays based on Michelson interfeometer sensor elements

levels for the array were measured to be in the range -50 to -65 dB.

Naturally, Mach-Zehnder arrays are not the only possible system topologies which can be used. Arrays based on Michelson interferometers have, for instance, been proposed and demonstrated. Figure 18 shows two Michelson configurations based on discrete and non-discrete interferometer elements, which are analogous to the Mach-Zehnder systems of Figures 15.a and 15.c.

Other developments in the area of TDM interferometric arrays include systems based on the reflectometric system of Figure 14.a, but using low-reflectivity fiber Bragg gratings as the partial reflectors [10]

5.5 Code-division multiplexing

Spread spectrum (SS) and code division multiplexed (CDM) techniques [46] have been applied to a variety of communications applications, including optical fiber systems [47]. This type of signal processing has also been previously investigated for optical time-domain reflectometry (OTDR) based sensing [48], and more recently, has been proposed and tested as a means for multiplexing interferometric sensors [49]. In this work, the interrogating laser source is modulated using a pseudo-random bit sequence (PRBS) of length $2^m -1$ (maximal length sequence, or m-sequence), and correlation is used to provide synchronous detection to identify specific sensor positions. A delay equal to an integer multiple of the bit (or 'chip') period separate the sensors. The received signals from the array are then encoded by delayed versions of the PRBS, and correlation techniques can be used to extract the individual signals.

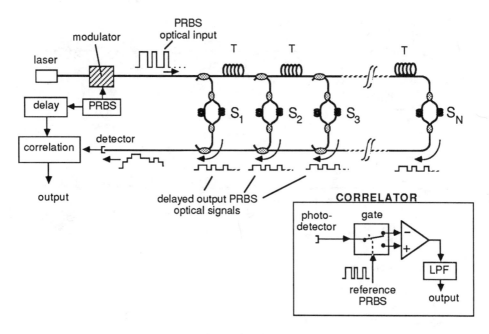

Figure 19. Code-division multiplexed array using Spread-Spectrum techniques

Figure 19 diagrammatically represents the principle of operation of the CDM approach applied to an interferometric sensor array. The PRBS input optical signal is fed to each the N sensors, delayed by a multiple, n_j, of the bit period T, where j denotes a specific sensor $(1 \leq j \leq N)$. The total output signal comprises the intensity sum of the overlapping delayed PBS sequences (each modified by the appropriate sensor transfer function). This results in a complex up-down staircase-like function at the optical detector which can be decoded using synchronous correlation-detection involving multiplication of the received signal with an appropriately delayed reference PRBS.

Although this method may provide advantages in terms of power budget for time-division multiplexed systems, it would also seem to be limited by excess phase noise effects arising due to mixing of time co-incident pulses from different sensors, and relatively high crosstalk between sensors. Recent work addressed these limitations of the technique, using a detection/signal processing approach which yields improved crosstalk and noise performance [50]. In this work, crosstalk levels lower than those expected from consideration of the code length were obtained using a mix of bipolar and unipolar codes which produces an improvement in the channel/channel isolation. This arises due to the correlation function of a bipolar with unipolar m-sequence PRBS, shown in Figure 20, which has a value $2^{(m-1)}$ for an aligned code, but is zero for any asynchronous alignment of the codes (this is in contrast to the conventional bipolar-bipolar auto-correlation which has a value of $(2^m -1)$ for code alignment, but a value of -1 for asynchronous alignment). This feature ensures good crosstalk can be obtained without the need to utilize excessively long PRBS codes: indeed, low crosstalk can be obtained providing the code length $(2^m -1) \geq N$, where N is the number of

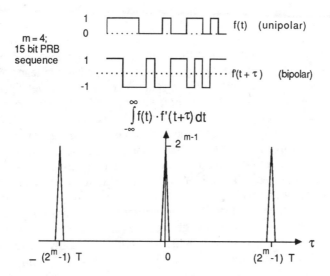

Figure 20. Correlation function between a bipolar and uniploar PRBS of length 2^m-1, with m = 4.

sensors in the array (assuming a one-bit time delay between sensors). Reduction of the excess phase noise by modulation of the laser source was also demonstrated in this work [50]

5.6 Power budget

Analysis shows that the power budget is a major factor determining the number of sensors which may be multiplexed using time-division and frequency-division schemes [51]. Calculations based on array loss models and input optical powers of 10 mW suggest that ~ 25 to 30 sensors may be supported per laser source with shot-noise equivalent phase detection sensitivities ~ 3 to 10 μrad./$\sqrt{Hz}$. This limitation to the multiplexing gain (number of sensors supported per input/output fiber) is due primarily to the relatively severe power recombinational losses associated with the use of conventional singlemode directional couplers in star or branching configurations. Generally, in a multiplexed fiber sensor array based on discrete sensor elements, light from a source is equally divided into the N sensors. On recombination of the sensor outputs onto a single monomode fiber, the optical throughput per channel is 1/N (effective loss of -10log[N] dB). A further effective power reduction factor of 1/N is encountered in time-division multiplexed (TDM) systems due to the duty-cycle of the pulsed source. In frequency division multiplexed (FDM) systems, a similar deleterious effect is encountered due to the fact that each channel is measured against a background of (N-1) other channels.

A novel singlemode/multimode (S/M) optical power combiner in multiplexed fiber sensor applications [52] has been demonstrated for improved power budget performance. This device allows a number of sensor outputs on single mode fibers to be efficiently recombined onto a single multimode output fiber with minimal effective insertion and excess loss.

5.7 Hybrid TDM/WDM system

An alternative means of improving the multiplexing gain is to utilize a system based on a hybrid of addressing approaches. A possible means for this is combining time or frequency division addressing with wavelength division multiplexing. Wavelength division multiplexing (WDM) has primarily been considered for use in communications systems. Its extension to the field of fiber sensors is obvious, and in principle the technique has the capability to allow a number of sensors to be remotely addressed in a very efficient manner. However, as discussed earlier, due to the complexity of the components, i.e.tree- or star-type WDM-couplers, required to selectively tap certain wavelengths from a fiber bus to sensors and recombine them onto a single output fiber, this approach has received little experimental attention. Furthermore, the crosstalk between sensors will be determined directly by the degree of wavelength isolation which can be achieved with the WDM-couplers, which is typically only ~ 15 to 20 dB. Consequently, WDM may not prove to be viable for the multiplexing of a significant number of sensors. However, combining wavelength division multiplexing concepts with time or frequency division addressing has the potential to allow a several-fold improvement in the number of multiplexed sensors in an array.

This type of system has been experimentally demonstrated [53]. The system, which is shown schematically in Figure 21, is based on two time division multiplexed systems which are addressed via common input and output fibers using wavelength division multiplexing to produce an array of 14 sensor elements. The source wavelengths are 835 nm and 790 nm, and the fiber WDM couplers used were manufactured by Aster Inc. The two sub-arrays comprised a ten-sensor system operating at 835 nm, and a four-sensor system designed for operation at 790 nm.

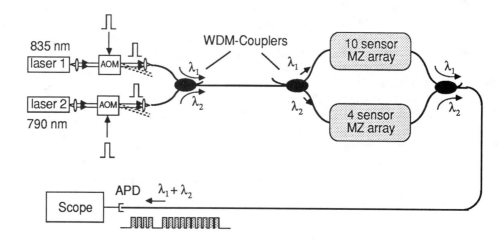

Figure 21. Hybrid TDM/WDM array

Both sources

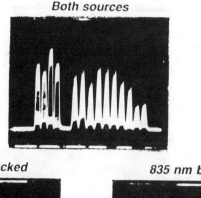

790 nm blocked

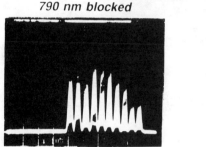

835 nm blocked

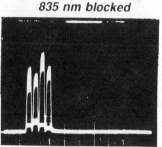

Figure 22. Outputs from the array configuration of Figure 20

Figure 21 shows the output pulse train observed with the system. Here, the top trace shows the detector (APD) output with both sources in operation. To provide a clearer visualization of the operation of the scheme, in this demonstration the input pulses from the two lasers were delayed relative to each other in order to separate in time the pulse trains produced by the two sub-arrays. In the two lower traces in Figure 21, the 835 nm and 790 nm lasers were blocked in turn to show the correct routing of the wavelengths in the system. Under normal operation, further wavelength de-multiplexing of the signals prior to detection would be utilized. Optical crosstalk levels < -25 dB (equivalent to < -50 dB electrical) between the two arrays were achieved with the components used.

5.8 Polarization fading

In recent years, considerable effort has been directed towards the development of techniques which provide compensation for the effects of polarization fading in interferometric sensor systems. Techniques based on the selection of a particular output state of polarization (SOP) or set of SOPs [54,55] have been demonstrated, which perform with limited success.

Very recently, a birefringence compensation approach has been used to develop a polarization independent Michelson interferometer [56]. An array based on this concept has also be demonstrated [57]. The birefringence compensation method is based on use of the "orthoconjugate reflector" of Edge and Stewart [58] which consists of a 45° Faraday rotator followed by a plane mirror. For an optical beam which retraces its path in a fiber, Pistoni and Martinelli [59] demonstrated that the insertion of a Faraday rotator and mirror (FRM) results in a state of polarization (SOP) at the exit which is orthogonal to the SOP at the input to the fiber. As

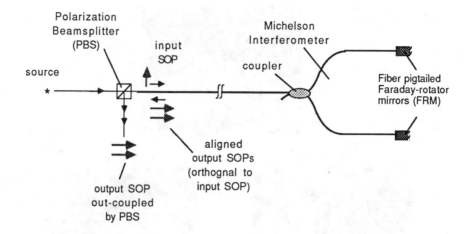

Figure 23. Polarization independent Michelson interferometer configuration

depicted in Figure 23, when employed in a Michelson interferometer the SOP from each of the device is returned orthogonal to the common input SOP. Consequently they are aligned with each other and insure maximum fringe visibility is obtained at the interferometer output. A four-sensor array configuration based on the configuration of Figure 18.b with miniaturized pigtailed FRMs as the reflectors has been built and tested. The fringe visibility was estimated to be > 0.95 (fading < 0.5 dB) simultaneously for all of the sensors under birefringence perturbations induced manually in the fiber leads [60].

6. TRANSDUCER MULTIPLEXING

The multiplexing techniques described in the foregoing section involve the networking of interferometric sensors. It is also possible to multiplex the transducer elements within a single interferometric sensor. Several fiber transducers have been developed in which the strain imparted to a fiber in an interferometer is proportional to the square to the applied measurand field. Examples include magnetostrictive [61] and electrostrictive materials [62], and a displacement sensing geometry [63] based on the lateral displacement of a fiber supported at two fixed points.

In general, the fiber strain can be expressed as

$$\varepsilon = C M^2,\tag{3}$$

where C is a constant which depends on the material parameters, or the exact geometrical arrangement of the transducer, and M is the measurand field (i.e. H, E or z in the cases of magnetic, electric fields and displacement respectively). If M comprises two components , $M_o + \Delta M \sin \omega_d t$, where M_o is proportional to the measurand field amplitude, and $\Delta M \sin \omega_d t$ is a 'dither' signal, the component of

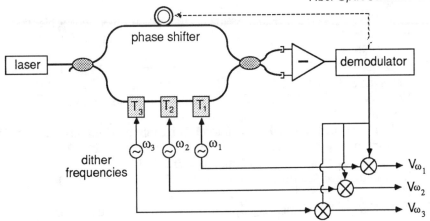

Figure 24. Interferometer configuration with transducer multiplexing

the strain induced in the fiber at the fundamental (ω_d) of the dither is

$$\varepsilon\,(\omega_d)\;=\;2C\Delta M\,M_o, \tag{4}$$

which is linearly proportional to M_o and thus the measurand field of interest. This strain can be detected using a fiber interferometer, and a number of such non-linear transduction elements can be incorporated in a single interferometer system by using different dither frequencies for each sensor, as shown schematically in Figure 24. Using this basic concept, the multiplexing of transducers for pressure, displacement and magnetic field [64] using a single interferometer has been demonstrated. Other measurands such as acceleration, and remote optical dithering have been demonstrated [65].

7. CONCLUSIONS

Work in the area of multiplexed fiber optic sensors has been reviewed. A key area of interest has been in the development of multiplexing techniques for high performance interferometric fiber sensor arrays, and the detailed coverage in this paper reflects this interest.

8. REFERENCES

1. A. D. Kersey and A. Dandridge," Distributed and Multiplexed Fiber Optic Sensor Systems," J. IERE, 58, p. S99, 1988.
2. J. P. Dakin, "Distributed fiber optic sensors", these proceedings.
3. E. Theochorous, " Distributed sensors based on differential absorption," Proc. of the IEE Colloquium on Distributed Optical Fiber Sensors (Digest 1986/74), paper #13, London, 1986.
4. S. A. Kingsley," Distributed Fiber-Optic Sensors: an Overview," Proc. SPIE Vol. 566, Fiber Optic and Laser Sensors III, p. 234, San Diego, CA, 1985.
5. A. J. Rogers," Intrinsic and Extrinsic Distributed Optical-Fiber Sensors," Proc. SPIE Vol. 566, Fiber Optic and Laser Sensors III, p. 234, 1985.

6. F. X. Desforges *et. al.*," Progress in OTDR Optical Fiber Sensor Networks," Proc SPIE Vol. 718, Fiber Optic and Laser Sensors IV, p. 225, Cambridge, MA, 1986.

7. G. Meltz et al., " Formation of Bragg Gratings in Optical Fiber by a Transverse Holographic Method", Optics Lett., 14,

8. W. W. Morey et al., " Bragg-Grating Temperature and Strain Sensors", Proc. OFS'89, p. 526, Paris, 1989 (Springer Verlag).

9. A. D. Kersey et al., " High Resolution Fiber Grating Based Strain Sensor With Interferometric Wavelength Shift Detection", Electron. Lett., 28, p. 236, 1992.

10. W. W. Morey et al., "Multiplexing Fiber Bragg Grating Sensors", Proc. 'Distributed and Multiplexed Fiber Sensors I', SPIE Proc. Vol. 1586, p. 216, 1991.

11. A. D. Kersey, " Multiplexing Options for Quasi-Distributed Sensing in Smart Structures Using Fiber Interferometry", Proc. Va Tech. Workshop on Optical Fiber Sensor Based Smart Materials and Structures, p. 6, Blacksburg, Va, April 1991 (Technomic).

12. A. R. Nelson *et. al.* ," Passive Multiplexing System for Fiber-Optic Sensors," Appl. Optics, 19, p. 2917, 1980.

13. W. B. Spillman and J. R. Lord," Self -Referencing Multiplexing Technique for Fiber-Optic Intensity Sensors," J. Lightwave Technol., LT-5, p. 865, 1987.

14. W. B. Spillman and J. R. Lord," Self -Referencing Multiplexing Technique for Fiber-Optic Intensity Sensors, SPIE Proc. Fiber Optic and Laser sensors V, San Diego, CA, 1987.

15. J. Mlodzianowski *et. al.*," A Simple Frequency Domain Multiplexing System for Optical Point Sensors," IEEE J. Lightwave Technol., LT-5, p.1002, 1987.

16. K. I. Mallalieu *et. al.*," FMCW of Optical Source Envelope Modulation for Passive Multiplexing of Frequency-Based Fiber-Optic Sensors," Electron. Lett., 22, p.809, 1986.

17. K. Fassgaenger et. al. "4 x 560 MBit/s WDM System using 3 Wavelength Selective Fused Single-mode Fiber Couplers as Multiplexer" Proc. ECOC '86, Barcelonia, Spain, Sept 1986.

18. H. Ishio, J. Minowa and K. Nosu, "Review and Status of Wavelength-Division Multiplexing Technology and its Application", J. Lightwave Technology, LT-2, pp. 448-463, 1984.

19. G. Winzer, "Wavelength Multiplexing Components - A Review of Single-mode Devices and their Applications", J. Lightwave Technology, LT-2, pp. 369–378, 1984.

20. C. A. Wade et al., " Optical Fiber Displacement Sensor Based on Electrical Subcarrier Interferometry Using a Mach-Zehnder Configuration", Proc. 'Fiber Optic Sensors', SPIE Conf. Proc Vol. 586, p. 223, Cannes, 1985.

21. C. A. Wade, A. D. Kersey and A. Dandridge, "Temperature Sensor Based on a Fiber-Optic Differential Delay RF Filter", Electron. Lett., 24, p.1305, 1988.

22. C. A. Wade, M. J. Marrone, A. D. Kersey and A. Dandridge, "Multiplexing of Sensors Based on Fiber-Optic Differential Delay RF Filters", Electron. Lett., 24, p.1557, 1988.

23. M. J. Marrone et al., " Quasi-Distributed Fiber optic Sensor System with Subcarrier Filtering", Proc. OFS'89, p. 519, Paris, 1989 (Springer Verlag).

24. A. D. Kersey, "Recent Progress in Interferometric Fiber Sensor Technology", Proc. 'Fiber Optic and Laser Sensors VIII', SPIE Proc. Vol. 1367, p. 2, 1990.

25. A.D. Kersey and A. Dandridge, "Multiplexed Mach-Zehnder Ladder Array with Ten Sensor Elements, Electron. Lett. 25, p.1298, 1989.

26. I.P. Giles, D. Uttam, B. Culshaw and D.E.N. Davies, "Coherent Optical-Fiber Sensors with Modulated Laser Sources", Electron. Lett., 19, p. 14, 1983.

27. S. Al Chalabi *et. al.* ," Multiplexed Optical Interferometers - An Analysis Based on Radar Systems," Proc. IEE Part J, 132, p. 150, 1985.

28. I. Sakai *et. al.*," Multiplexing of Optical Fiber Sensors Using a Frequency-Modulated Source and Gated Output," IEEE J. Lightwave Technol., LT-5, p.932, 1987.

29. A. Dandridge, A.B. Tveten and T.G. Giallorenzi, "Homodyne Demodulation Scheme for Fiber-Optic Sensor Using Phase Generated Carrier", IEEE J. Quantum Electron., 18, p. 647, 1982.

30. A. Dandridge, A.B. Tveten, A.D. Kersey and A.M. Yurek, "Multiplexing of Interferometric Sensors using Phase Generated Carrier Techniques", IEEE J. Lightwave Technology, LT-5, p. 947, 1987.

31 A. Dandridge et al., "AOTA Tow test results", Proc. AFCEA/DoD Conf. on Fiber Optics '90, p. 104, McLean, 1990.

32. S.A. Al -Chalabi, B. Culshaw, and D.E.N. Davies, "Partially Coherent Sources in Interferometric Sensors",Proceedings of the First International Conference on Optical Fibre Sensors (IEE), pp. 132-135, 1983.

33. J. L. Brooks e. al., " Coherence Multiplexing of Fiber Optic Interferometric Sensors", IEEE J. Lightwave Technol., 3, p. 1062, 1985.

34. A.D. Kersey and A. Dandridge, "Phase Noise Reduction in Coherence Multiplexed Interferometric Fiber sensors", Electron. Lett., Vol. 22, No. 11, pp. 616-618, 1986.

35. D. O'Connell, A. D. Kersey, A. Dandridge, and C. A. Wade, "Coherence Multiplexed Fiber Optic Temperature Sensor using a Wavelength Dithered Source", Proc. OFC '89, p. 145, Houston, Feb. 1989. (OSA)

36. A. D. Kersey et al., " Differential polarimetric Fiber Optic Sensor Configuration with Dual Wavelength Operation", Appl. Optics, 28, p. 204, 1989.

37. V. Gusmeroli et al., " A Coherence Multiplexed Quasi-Distributed Polarimetric Sensor Suitable for Structural Monitoring", Proc. OFS'89, p. 513, Paris, 1989 (Springer Verlag).

38. J.P. Dakin, C.A. Wade and M.L. Henning, "Novel Optical Fibre Hydrophone Array Using a Single Laser Source and Detector", Electron. Lett., 20, p. 53, 1984.

39. M.L. Henning and C. Lamb, "At-Sea Deployment of a Multiplexed Fiber Optic Hydrophone Array," Proc. 5 th Int. Conf. on Optical Fiber Sensors, p. 84, New Orleans, Jan. 1988 (OSA technical digest).

40. A. D. Kersey, K. L. Dorsey and A. Dandridge, "Demonstration of an eight-element Time-Division Multiplexed Interferometric Fiber Sensor Array," Electron. Lett., 24, p. 689, 1988.

41. A. D. Kersey, A. Dandridge and K. L. Dorsey, "Transmissive Serial Interferometric Fiber Sensor Array," J. Lightwave Technol., 7, p. 846, 1989.

42. A. D. Kersey et al., " Analysis of Intrinsic Crosstalk in Tapped Serial and Fabry-Perot Interferometric Fiber Sensor Arrays",, Proc. 'Fiber Optic and Laser Sensors VI', SPIE Proc. Vol. 985, p. 113, Boston, 1988.

43. B. Moslehi et al., "Efficient Fiber Optic structure with Applications to Sensor Arrays, IEEE J. Lightwave Technol., 7, p. 236, 1989.

44. Brooks et al.," Fiber Optic Interferometric Sensor Arrays with Freedom From Source Induced Phase Noise", Opt. Lett., p.473, 1986.

45. A.D. Kersey, A. Dandridge and A.B. Tveten,"Multiplexing of Interferometric Fiber Sensors Using Time Division Addressing and Phase Generated Carrier Demodulation", Optics Letters, 12, p. 775, 1987.

46. R. C. Dixon, "Spread Spectrum Systems", Wiley, 1984.

47. P. R. Prucnal et al., "Spread Spectrum Fiber Optic Local Area Network using Optical Processing", J. Lightwave Technol., LT-4, p. 547, 1986.

48. J. K. A. Everard, " Novel Signal Processing Techniques for Enhanced OTDR Sensors", Proc. Fiber Optic Sensors II, SPIE vol. 798, p. 42, The Hague, 1987.

49. H. S. Al-Raweshidy and D. Uttamchandani, " Spread Spectrum Technique for Passive Multiplexing of Interferometric Fiber Optic Sensors", Proc. Fiber Optics'90, SPIE vol. 1314, p. 342, London, 1990.

50. A.D. Kersey and A. Dandridge, "Low Crosstalk Code division Multiplexed Interferometric Array", Electron. Lett., 28, p.351, 1992.

51. A. D. Kersey and A. Dandridge, "Comparative Analysis of Multiplexing Techniques for Interferometric Fiber Sensors", Proc. 'Fiber Optics 89, SPIE Conf. Proc. Vol 1120, p. 236, London, 1989

52. A. Dandridge et. al., " Increasing Multiplexed Fiber Sensor Array Performance by Use of a Singlemode/Multimode Recombiner", Proc. 6th In. Conf. on Optical Fiber Sensors, OFS'89, Post deadline paper # 5, p. 40, 1989.

53. A.D. Kersey and A. Dandridge, "Demonstration of a Hybrid Time/Wavelength Division Multiplexed Interferometric Fiber Sensor Array", Electron. Lett., 2, p.554, 1991.

54. N. J. Frigo, A. Dandridge and A. B. Tveten, "Technique for elimination of polarization fading in fiber interferometers," Electron. Lett., 20, p. 319, 1984.

55. A. D. Kersey, M. J. Marrone, A. Dandridge and A. B. Tveten, "Optimization and Stabilization of Visibility in Interferometric Fiber-Optic Sensors Using Input-Polarization Control", J. Lightwave Technol. 6, p. 1599, 1988.

56. A. D. Kersey, M. J. Marrone and M. A. Davis, "Polarization-Insensitive Fiber Optic Michelson Interferometer", Electron. Lett. 26, p. 518, 1991.

57. M. J. Marrone et al., "Fiber Michelson Array with Passive Elimination of Polarization Fading and Source Feedback Isolation", Proc. OFS'92, p. 69, Monterey, 1992 (IEEE).

58. C. Edge and W. J. Stewart, "Measurement of Nonreciprocity in Single-Mode Optical Fibers", Tech. Dig. IEE Colloq. on Optical Fiber Measurements, No. 1987/55, 1987.

59. N. C. Pistoni and M. Martinelli, "Birefringence Effects Suppression in Optical Fiber Sensor Circuits", Proc. 7th Int. Conf. on Optical Fiber Sensors, p. 125, 1990.

60. M. J. Marrone et al., "Polarization Independent Interferometric Array Configurations", Proc. 'Distributed and Multiplexed Fiber Optic Sensors II", SPIE vol 1797, 1992.

61. A. D. Kersey et al., "Detection of DC and Low Frequency AC Magnetic Fields Using an All Single-Mode Fiber Magnetometer", Electron. Lett., 19, p. 469, 1983., and K. P. Koo et al., "A Fiber Optic DC Magnetometer, IEEE J. Lightwave Technol., LT-1, p. 524, 1983.

62. S. T. Vohra et al., "Fiber Optic DC and Low Frequency Electric Field Sensor", Optics Lett., 16, p. 1445, 1991.

63. A. D. Kersey et al., " New Nonlinear Phase Transduction Method for DC Measurand Interferometric Fiber Sensors", Electron. Lett., 22, p. 75, 1986.

64. F. Bucholtz, A. D. Kersey and A. Dandridge, " Multiplexing of Nonlinear Fiber Optic Interferometric Sensors", IEE J. Lightwave Technol., 7, p. 514, 1989.

65. A. D. Kersey, F. Bucholtz, K. Sinansky and A. Dandridge," Interferometric Sensors for DC Measurands - A New Class of Fiber Sensors," SPIE Proc. vol. 718., ' Fiber Optic and Laser Sensors IV', Cambridge, MA, p. 198, 1986.

SESSION 5

Applications of Fiber Optic Sensors

Chair
William B. Spillman, Jr.
University of Vermont

Recent advances in fiber optic magnetic sensing

F. Bucholtz

Naval Research Laboratory,
Optical Sciences Division, Code 5670
Washington, DC 20375-5000

ABSTRACT

Considerable progress has been made in the use of fiber optic magnetic sensors in actual field environments, especially for undersea applications. This paper reviews the results of three fiber magnetometer systems: 1) a heading sensor for undersea towed array applications; 2) a remote ac magnetometer designed for land use; and 3) a magnetometer array for undersea magnetic measurements. In each case, relevant design parameters, laboratory test results, and, where applicable, field test results are presented.

1. INTRODUCTION

Fiber optic magnetometers based on magnetostriction have been under development for approximately ten years. Thus far, early projections of the performance of fiber magnetometers [1] have been achieved only at audio frequencies [2] while at low frequencies (< 10 Hz), the resolution is limited by a variety of upconversion processes, both optical and magnetic. Laboratory devices have demonstrated 10 pT / $\sqrt{Hz}$ resolution at 1 Hz [3], a value roughly equivalent to high performance fluxgate magnetometers. Hence, the viability of fiber magnetometers in the commercial and military marketplace cannot be based solely on raw performance for single devices. Fiber magnetometers will succeed based on their ability to be integrated into sensor systems, especially systems which operate over long distances or which require the measurement of multiple parameters, or both. In these cases, the multiplexing capabilities and low attenuation properties of fiber optic systems present clear advantages over competing technologies. Development of the fiber magnetometer has reached a stage where transducer engineering and electronics are the significant issues for field deployable systems. Issues related to interferometer operation have been largely solved.

This review paper emphasizes recent work at NRL on magnetostriction-based fiber optic magnetometers. The focus is on field tested systems although new results obtained from laboratory systems are also presented. Additional work performed at other laboratories is summarized. SI units are used throughout with magnetic field measured in Tesla (T) where $1 \text{ T} = 10^4$ Gauss. A convenient unit is the nanoTesla ($1 \text{nT} = 10^{-5}$ G).

Previous review papers summarized related work through mid - 1990 [4,5]. Excellent review papers on magnetometry in general can be found in the literature [6,7].

Various methods exist for the detection of magnetic fields using fiber optics including 1) devices based on the Faraday effect [8,9], in which a polarimeter measures the change in the state of polarization caused by a magnetic field; and 2) devices based on magnetostriction, in which an interferometer measures the change in optical phase caused by the change in physical length of a magnetically-sensitive material. Due to time reversal symmetry and the absence of magnetic monopoles, magnetostriction in materials must depend only on even powers of the magnetic field. To lowest order the strain $e = \Delta L / L$ is related to total applied field H by

$$e = CH^2, \tag{1}$$

where L is the sample length, ΔL is the change in length, and the magnetostriction parameter C is a function of extrinsic transducer parameters (boundary conditions, mechanical loading, and demagnetization) and intrinsic material parameters (saturation magnetization and saturation magnetostriction). Spatial variations in C have been ignored for this discussion. For materials such as iron-boron metallic glass alloys typically used in transducers, $C \approx 10^2 \, T^{-2}$.

If the total field H consists of a static field H_o and an oscillating field $h\sin\omega t$, then the strain at frequency ω is given by $e(\omega) = 2CH_o h\sin\omega t$. Hence, the transducer can operate as either a dc magnetometer, by applying a known oscillating field, or as an ac magnetometer by applying a known dc magnetic field. Both types of magnetometer are discussed in this review.

Fluctuations in the strain due to thermal energy kT represent a fundamental limit to the resolution of magnetostriction-based magnetometers. The magnitude of this noise can be estimated using the fluctuation-dissipation theorem [10] which relates the mean square strain fluctuation spectrum $<e^2(\omega)>$ to the imaginary (dissipative) part of the elastic compliance $G_{em}"(\omega)$,

$$<e^2(\omega)> = 2kT \mid G_{em}"(\omega) / \omega \mid / EV\pi, \tag{2}$$

where E and V are the sample's Young's modulus and volume, respectively. Figure 1 shows the thermal strain noise spectrum of a cylindrical metallic glass transducer [11]. It was the first demonstration of thermal-noise-limited performance in a fiber optic transducer. The noise peak just below 26 kHz is equivalent to approximately 0.8 μrad / $\sqrt{Hz}$ and, for this transducer, corresponds to approximately 100 fT / $\sqrt{Hz}$ minimum detectable ac magnetic field. The magnetic field equivalent noise due to thermal strain fluctuations depends on temperature, frequency, and transducer parameters. In general, different transducers will exhibit completely different thermal-strain-noise limited performance.

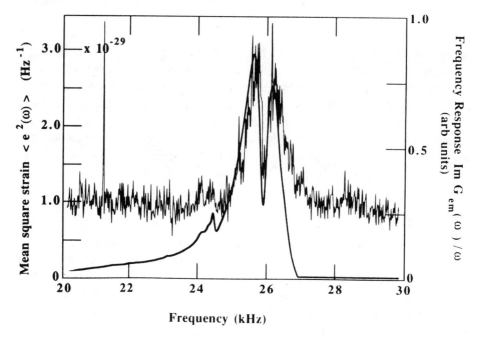

Fig. 1. Thermal strain spectrum of a cylindrical metallic glass cylinder compared to the prediction of the fluctuation-dissipation theorem based on the imaginary part of the elastic compliance $G_{em}''(\omega)/\omega$ (bold line).

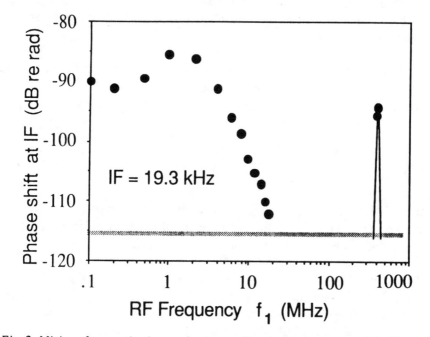

Fig. 2. Mixing of magnetization modes in metallic glass strip measured by fiber interferometry. The roll-off near 3 MHz is due to eddy-current damping and the response near $f_1 = 400$ MHz is attributed to spin wave modes.

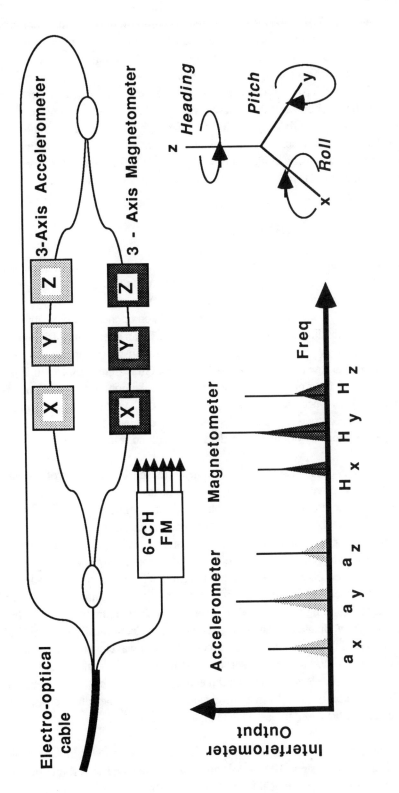

Fig. 3 . Signal processing for the fiber optic magnetic heading sensor. Six dithers are sent to the sensor over a single fiber using a 6-channel analog FM link. The orientation angles heading, pitch and roll are defined in the lower right inset.

Although the relation Eq.(1), the so called "Coherent Rotation Model," is sufficient to account for a significant fraction of the observed magnetostriction in amorphous alloys, certain experiments have demonstrated the invalidity of Eq. (1). Figure 2 shows the results of an experiment in which a magnetostrictive element was excited by two ac magnetic fields $h_1 \sin 2\pi f_1 t$ and $h_2 \sin 2\pi f_2 t$ such that the difference $f_2 - f_1 = IF$ was kept constant [12]. The figure shows that the material was capable of mixing spin wave modes due to applied magnetic fields near 400 MHz and suggests that a more suitable expression for the magnetostriction may involve the square of the total magnetization M or the square of the total magnetic induction $B = \mu_0(H + M)$. Nevertheless, for most transducer applications, Eq. (1) represents a valid working model.

2. HEADING SENSOR

Navigation is an important application for magnetic sensors. A magnetic heading sensor works by measuring the ratio of the two horizontal components of the Earth's magnetic field, H_x and H_y, to obtain the heading angle θ with respect to magnetic north, $\theta = \arctan(Hy / Hx)$. Measurement of the horizontal field components is typically assured in practice by gimballing a two-axis magnetometer with an attendant restriction in the permissible pitch angle. Since gimballing was not a realistic option for the fiber magnetometer, the heading sensor developed at NRL used a miniature three-axis magnetometer and a three-axis fiber optic accelerometer based on the nonlinear displacement-to-strain technique [13]. Once the the pitch ψ and roll ρ of the sensor are determined from the accelerometer outputs, magnetic heading can be obtained from the three magnetometer outputs for arbitrary orientation of the sensor with no restriction on ψ or ρ.

The three magnetic transducers and three accelerometer transducers were part of a single interferometer (Fig. 3) driven by a single semiconductor laser at 0.83 μm wavelength. Design of the fiber optic accelerometers was based on the nonlinear displacement -to-strain technique. One dither signal is required for each transducer, six dithers in all. The laser, interferometer and magnetometer demodulators, and signal recording equipment are located at the ship end of the electrooptical cable and are not shown in Fig. 3. The sensor requires three fibers for operation: interferometer input and output leads, and one additional fiber for transmitting the six dither signals. Dither signals were transmitted to the sensor using an analog FM scheme. Six carriers in the 10 - 12 MHz range, each frequency modulated by one of the dithers, were electronically summed and the summed signal modulated the current of a second 0.83 μm wavelength semiconductor laser. An electronics module near the sensor demodulated the optical intensity-modulated FM carriers into six electrical signals: three voltage dithers for the piezoelectric elements in the accelerometers and three current dithers for the magnetic transducers. Phase generated carrier was used to demodulate the interferometer.

The magnetic transducer consisted of a single strip of field-annealed [14] metallic glass (Metglas 2605S-2) of dimensions 10 mm x 1 mm x 25 μm bonded to a single pass of jacketed, communications-grade optical fiber (York SM800). Each dither coil consisted of 120 turns of 40 AWG wire wound on a brass mandrel of dimensions 15 mm length x 3.0 mm diameter. The three dither coils were potted into orthogonal holes machined in a 1 cm^3 phenolic cube. A typical frequency response for the magnetic transducers is shown in Fig. 4. The response has a Lorentzian lineshape characteristic of a driven, linear mechanical oscillator with damping. Operation of transducers dithered both at and below the resonance frequency was investigated.

The design of the fiber optic accelerometers was based on the nonlinear strain-to-displacement technique [13]. A brass proof mass was mounted at the end of a brass cantilever arm. Mounted on the proof mass was a small PZT bimorph element attached to the fiber at the midpoint between two support points, a distance 2D apart, to provide the displacement dither $z(\omega) = z \sin\omega t$. The quasistatic displacement z_0 of the proof mass is a function of the angle α between the sensor and the vertical, $z_0 = mg\cos\alpha / k_{eff}$, where m is the mass of the proof mass, $g = 9.8$ m / sec^2 is the acceleration due to gravity, and k_{eff} is the effective spring constant of the cantilever arm. In transducers of this type it is important that the magnitude of both the static and dither displacements remain small compared to the fiber length 2D in order to preserve the quadratic relationship between total displacement $z = z_0 + z(\omega)$ and longitudinal strain e in the fiber, $e = z^2/2D$, and in order to avoid excessive stretching of the fiber jacket, which can lead to irreproducibility in the scale factor.

Transformation of the sensor's six outputs to the desired orientation angles heading, pitch, and roll is accomplished by use of orthogonal rotation matrices. The sensor measures, in its own reference frame, an acceleration vector $\mathbf{a} = [ax, ay, az]$ and a magnetic field vector $\mathbf{H} = [Hx, Hy, Hz]$. By definition, in the Earth's magnetic North reference frame, $\mathbf{a_e} = [0, 0, g]$ and $\mathbf{H_e} = [H\cos I, 0, H\sin I]$. The z axis is positive downward, H is the magnitude of the Earth's magnetic field, and I is the magnetic inclination angle. Vectors in the sensor frame and the Earth frame are related by a transformation matrix M which itself is the product of three orthogonal matrices corresponding to heading θ, pitch ψ, and roll ρ, $M = M_\rho M_\psi M_\theta$. Hence, $\mathbf{a} = M\mathbf{a_e}$ and $\mathbf{H} = M\mathbf{H_e}$. The individual matrices M_ρ, M_ψ, and M_θ are exactly those known as the Eulerian angles for coordinate transformation [15]. All <u>four</u> angles θ, ψ, ρ, and I may be obtained by performing calculations on the six outputs a_x, a_y, a_z, H_x, H_y, and H_z [16]. These calculations may be done either in dedicated hardware or, as was done for the sensor reported here, off-line using software.

Table I summarizes the performance of the heading sensor under actual undersea operating conditions with the sensor towed approximately 300 m behind the shipboard signal processing electronics. Comparison of the fiber heading sensor output with a "conventional" gimballed two-axis fluxgate heading sensor in Fig. 5 shows good agreement obtained when the fiber magnetometers were dithered at a frequency well below resonance. The performance is good for a prototype device and improvements must be made only in the average heading error category. This error was attributable mainly to the nonideal behavior of the fiber optic accelerometers.

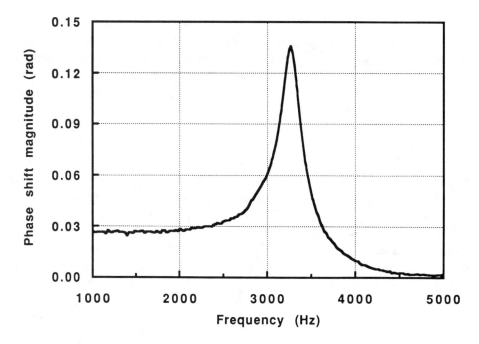

Fig. 4. Typical dither frequency response of magnetostrictive transducer for fiber optic heading sensor.

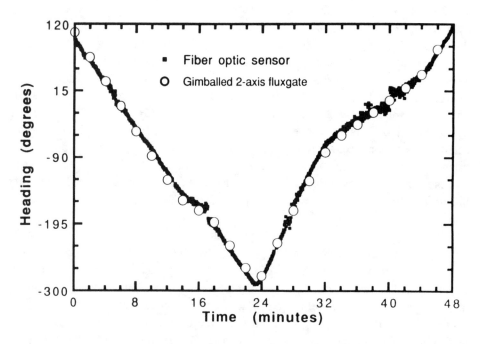

Fig. 5. Comparison of heading measured by fiber sensor and fluxgate sensor under actual at sea operating conditions.

TABLE I. Summary of fiber optic heading sensor performance under actual at sea operation.

Parameter	Value
Interferometer phase resolution (1-50 kHz)	$\leq 15\ \mu\mathrm{rad}/\sqrt{\mathrm{Hz}}$
Magnetic field resolution (dc - 10 Hz)	$\leq 500\ \mathrm{nT}/\sqrt{\mathrm{Hz}}$
Heading resolution(dc - 10 Hz)	≤ 12 arc minutes $/\sqrt{\mathrm{Hz}}$
Average heading error	≤ 3 deg
Orthogonality	≤ 0.5 deg

3. REMOTE AC MAGNETOMETER

An important frequency range for ac magnetometers is 1 - 100 kHz where signals exist from a variety of man-made devices. Important requirements for a practical device operating in this range are low power, good magnetic field resolution, small size, and immunity to changes in the external static magnetic field. A remotely-operated fiber optic ac magnetometer was developed to meet these requirements [17].

The system configuration is shown in Fig. 6. Active homodyne demodulation is used with feedback to the laser emission frequency. A compact, unannealed cylindrical metallic glass transducer, 12 mm diameter x 25 mm length x 25 μm thick was biased to approximately 1.5 mT using a small permanent magnet placed inside the cylinder. The sensitivity of the magnetometer response to external dc field, both with and without the bias field, is shown in the inset. Changes in the orientation of the sensor in the Earth's magnetic field will change the bias field by no more than ± 0.05 mT. Since the bias point of the sensor is chosen on the flat region just above the knee, the responsivity is virtually independent of dc field in this range.

Figures 7 and 8 show the sensor's vector response, and frequency response and resolution, respectively. These data were obtained in the laboratory. The frequency response is smooth and relatively flat , with no more than 6 dB amplitude rise, from 1 kHz to approximately 50 kHz. Again, the resonance lineshape is Lorentzian with the peak occurring at the first longitudinal mechanical resonance of the metallic glass cylinder. The interferometer-limited noise floor is equivalent to magnetic field resolution of approximately 18 pT $/\sqrt{\mathrm{Hz}}$, more than sufficient to meet most requirements. As a function of angle between the external field and the axis of the sensor, the data in Fig. 7 shows the expected cosine antenna pattern.

An additional factor to be considered with an ac magnetometer is harmonic distortion. This is especially critical with magnetostriction-based magnetometers since the transducing mechanism is intrinsically nonlinear. Independent of bias field, an ac magnetic

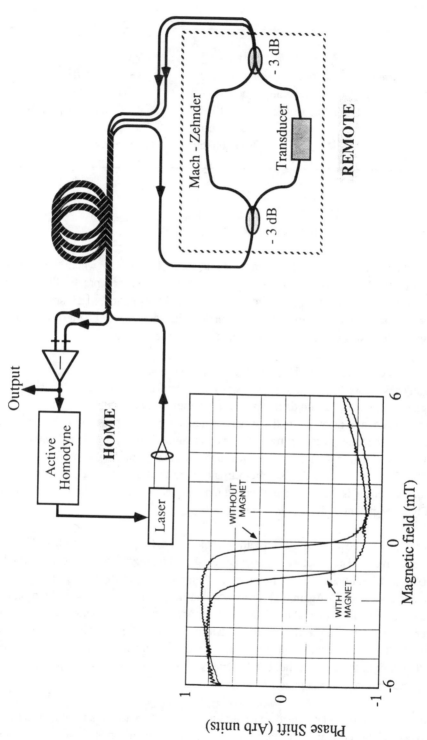

Fig. 6. Remote ac magnetometer system. Interferometer is demodulated by active homodyne feedback to the laser emission frequency. Inset shows dc field dependence of transducer signal both with and without bias magnet.

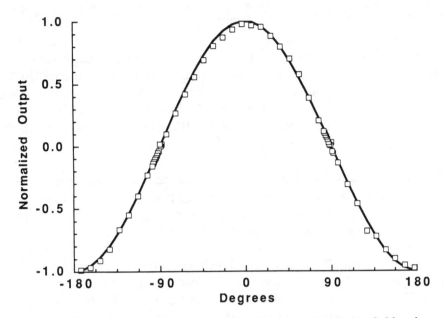

Fig. 7. *Remote ac sensor output as function of angle between applied ac field and sensor axis. Solid line is cosine curve.*

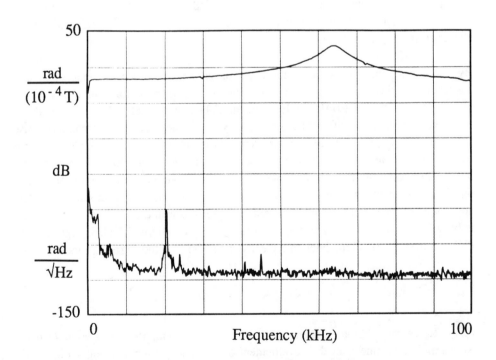

Fig. 8. *Frequency dependence of remote ac sensor responsivity (top trace) and noise floor (bottom trace). Noise equivalent magnetic field is approximately 20 pT / √Hz. Spurious signal near 20 kHz is due to test equipment.*

field $h\sin\omega t$ produces a strain component at 2ω with amplitude $(Ch^2/2)$. Defining harmonic distortion as the ratio of the amplitude of the second harmonic term $Ch^2/2$ to the fundamental term $2CH_oh$, the magnetostrictive transducer exhibits intrinsic second harmonic distortion $SHD = h/4H_o$. Increasing the bias field H_o to reduce SHD is effective only up to H_o values where the metallic glass sample begins to saturate. At fields significantly larger than the ones considered in the inset in Fig.6, the response at the fundamental decreases, as well . For the sensor described here, $Ho = 1.5$ mT and the distortion $SHD = (1.7\times10^{-7})h(nT)$. Hence the SHD remains below 0.1 % for $h<5800$ nT (58 mG).

The magnetic field resolution shown in Fig. 8 is limited entirely by the phase resolution of the interferometer which was approximately 0.3 μrad / $\sqrt{Hz}$ in the frequency range shown. The best reported ac resolution for a fiber magnetometer is 70 fT / $\sqrt{Hz}$, obtained under somewhat contrived conditions by operating directly at a mechanical resonance near 34 kHz[2]. Resolution in this case was again limited by the phase resolution of the interferometer although the thermal strain noise floor was within 6 dB of the interferometer noise floor. In general, the reported ac resolution for magnetostriction-based fiber magnetometers has been limited by either the interferometer or by the thermal strain noise of the transducer. (It is instructive to note that the interferometer-limited 70 fT / $\sqrt{Hz}$ resolution discussed here is better than the thermal-limited 100 fT / $\sqrt{Hz}$ resolution discussed in the Introduction. The difference, which is attributable to differences in parameters of the transducers - most importantly fiber length and C value - illustrates that temperature alone does not determine the thermal-noise-limited resolution of a magnetostrictive transducer.

By comparison, the low-frequency resolution of magnetostriction-based magnetometers has reached neither the interferometer noise limit nor the thermal noise limit. For example, the 10 pT / $\sqrt{Hz}$ at 1.0 Hz resolution reported by Dagenais et.al. [3] was well above the thermal noise limit and was approximately 10 dB above the interferometer noise floor. In addition, the self-noise power typically increases with 1/f dependence at low frequencies - clear indication of a noise mechanism related to upconversion and not related to baseband interferometer or thermal noise. Although the sensing mechanism in the fiber magnetometer relies on the $e = CH^2$ relation in both dc and ac sensing modes, application of the dither field in dc operation of the sensor in the dc mode puts the metallic glass element in a dynamic magnetostrictive condition which is significantly different from the small modulation typically encountered in the ac sensor mode.

4. MAGNETIC ARRAY

Naval Research Laboratory is currently collaborating with the Norwegian Defence Research Establishment on the development of an undersea array of eight fiber optic vector magnetometers, a Magnetic Array System (MARS). The performance goals of the MARS system are summarized in Table II.

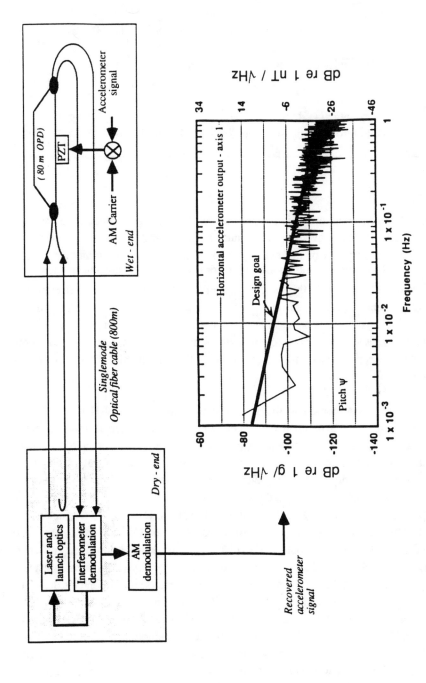

Fig. 9. System for 1991 MARS test for magnetometer platform stability showing accelerometer data telemetered to "dry-end" as a phase shift on the Mach-Zehnder interferometer (1.3 μm). Inset shows recovered low-frequency data for motion along the pitch axis and the equivalent worst-case magnetic field fluctuation (right axis). Design goal is shown.

TABLE II. Performance goals for the fiber optic Magnetic ARray System (MARS).

1. Magnetometer

Self noise	≤ 0.3 nT / $\sqrt{\text{Hz}}$ at 0.1 Hz
Crosstalk	< -80 dB
Sensitivity to orthogonal fields	< -80 dB
Dynamic range	≥ 80 dB

2. Electro-optical system

Interferometer phase resolution	≤ 1.0 µrad / $\sqrt{\text{Hz}}$

When using vector magnetometers in the presence of the Earth's magnetic field, stability of the sensor is needed to avoid motion-induced signals. Indeed, it has been suggested that a significant portion of the existing data on low-frequency ambient magnetic noise is corrupted by sensor motion noise [18]. For example, for a transducer initially aligned perpendicular to the Earth's field H_o, a 20 µrad (≈ 1.0 millidegree) decrease in the angle of the sensor with respect to H_o results in a 1.0 nT output - a value roughly equivalent to the rms ambient noise in the 1 mHz to 10 mHz frequency range. Since sensor stability is essential for low-frequency vector magnetic measurements, a preliminary MARS test was performed in 1991 to determine the stability and deployability of a nonmagnetic anchor.

The electro-optical system used in the 1991 test is shown in Fig. 9 [19]. A 800 m Alcatel slotted core, electro-optical cable containing 10 optical fibers (singlemode, 1.3 µm wavelength) and 16 copper wires was used. All data was telemetered to shore over the interferometer output leads; the wires were used only to deliver dc electrical power to electronics in the sensor. The sensor package, consisting of Mach-Zehnder interferometer, 3-axis accelerometer, and associated electronics was housed in a steel enclosure and deployed undersea at a depth of approximately 150 m. No magnetometer was tested. Dither signals used for telemetry (discussed below) were sent to the node over a dedicated fiber using an optical FM link similar to the one described earlier in Section 2 (Heading Sensor). In the shore-based station, the "dry-end" system consisted of diode-pumped Nd:YAG laser, launch optics, photodetectors, interferometer demodulator, signal processors and data recorders. In addition, since only one interferometer was used, it could be demodulated using either Phase Generated Carrier (PGC) or active homodyne locked to the laser emission frequency. A secondary purpose of the test was to evaluate the performance of these two demodulation schemes for low-frequency sensing. Motion of the sensor was measured using a three-axis accelerometer consisting of three Sundstrand QA 700 accelerometers. Low-frequency data from the accelerometers was impressed as phase shifts on the interferometer using a simple AM modulation scheme. Finally, two fiber leads in the cable were spliced together at the sensor forming a 1.6 km length used for polarization studies.

The inset in Fig. 9 shows the low-frequency power spectrum of the accelerometer x-axis, corresponding to pitch angle ψ, after deployment. The abscissa is scaled in dB relative to 1 g / $\sqrt{Hz}$ (i.e. 20 log (ax/g)) where, again, g = 9.8 m/sec^2. In the case of pitch, the angle ψ is determined simply from the accelerometer output as $\psi = \sin^{-1}(ax/g)$.

In the AM telemetry scheme, it was found that low-frequency data was susceptible to both low-frequency relative intensity noise (RIN) and demodulator noise. Low-frequency RIN of the light in the interferometer, due to intrinsic RIN of the laser and RIN introduced by the launch optics, mixes with the phase carriers used for AM telemetry and introduces noise with a fixed ratio to the carrier [20]. It was also found that signals associated with fringe cycling in the PGC demodulation could also appear in the sidebands of the phase carriers. This effect is entirely a function of the electronic balance in the sine/cosine portion of the demodulator. One method for reducing these two noise mechanisms is to use one phase carrier, containing no data, in order to recover a voltage containing only low-frequency noise. The noise can then be subtracted from each of the data channels. The data shown in Fig. 9 was obtained using this technique on stored data, no real time subtraction was attempted. Approximately 20 dB suppression of common mode noise suppression was achieved at frequencies below 10 mHz.

The spectrum of angular fluctuations was converted to equivalent motion-induced magnetic field fluctuations using $H = H_o \sin\psi$, where H_o = 50,000 nT is the magnitude of the Earth's field. The right hand abscissa in the inset, Fig.9, is scaled in units of 20 log H(nT)/$\sqrt{Hz}$). Also shown is the design goal. Hence, the stability of the anchor was sufficient to meet the design goal.

One novel feature of the system was the ability to monitor the sensors during deployment. Figures 10 shows the accelerometer outputs just before and just after the sensor reached the sea floor. The sensor remained relatively level while it was lowered through the water and came to rest at an orientation having angles 6.5 deg pitch and 15 deg roll (Fig. 10). The sensor remained at this orientation for the length of the test with angular fluctuations as given by the spectrum in Fig. 7.

5. RELATED WORK

Recently, a fiber optic magnetic gradiometer successfully measured the magnetocardiogram of a human heart [21]. Use of the gradiometer is essential in this application since ambient magnetic field fluctuations in the clinical settings are much stronger than the heart-induced signals and use of magnetic shielding is typically not practical. This result represents the first biomedical application of a fiber optic magnetostrictive sensor.

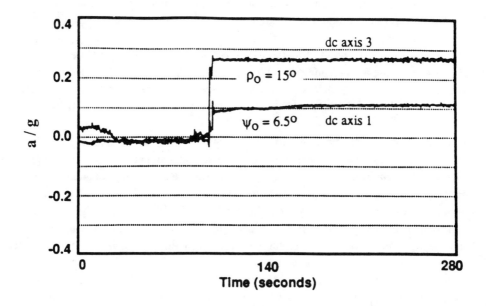

Fig. 10 System outputs during deployment of the MARS prototype sensor showing accelerometer outputs corresponding to pitch ψ and roll ρ. The sensor reaches the sea floor at approximately t = 100 sec.

A significant issue still under investigation is the residual signal in the dynamic magnetostriction [3]. This signal results from nonideal behavior of the material. That is, the material does not precisely follow Eq. 1. The residual signal allows the upconversion of low-frequency nonmagnetic noise and thereby degrades the signal-to-noise ratio of the sensor. Recently, a balanced detection electronic scheme has been proposed to suppress these effects [22]. The technique was successful in eliminating the residual signal from the final output but more measurements are needed to demonstrate improvement in resolution using this method.

6. SUMMARY

Recent work on fiber optic magnetic sensor based on magnetostriction has focussed on the design and testing of devices in the field, or, more accurately, in the sea. A fiber optic heading sensor was successfully deployed and tested and achieved heading resolution below 1 degree. A compact, remotely-operable ac magnetometer was tested in the laboratory and found to have excellent resolution (< 20 pT / $\sqrt{Hz}$) and vector response characteristics. Work is currently in progress on the development of an undersea fiber optic magnetometer array. Preliminary tests verified various performance characteristics of the mechanical and signal processing components of the system.

7. ACKNOWLEDGEMENTS

The author wishes to acknowledge significant contributions made by the following individuals in the work summarized in this review. James A. McVicker of SFA, Inc. was responsible for design and manufacturing of the magnetostrictive transducers used in all the NRL work reported. Kris G. Wathen, formerly of DSI, Inc., designed and constructed the 6-channel optical FM link for the heading sensor and the MARS 1991 test. Carl A. Villarruel (NRL) and Allen R. Davis of University Research Foundation of Maryland designed and built the dry-end optical system for the MARS test. In addition, Scott Patrick of Virginia Polytechnic and State University and A.R. Davis designed and tested the ac magnetometer. Scott Patrick was also responsible for signal processing, data acquisition and data analysis for the heading sensor. Dominique M. Dagenais and K. P. Koo contributed to the design and testing of the heading sensor. A. D. Kersey (NRL) suggested the orthogonal matrix approach for the heading sensor. A. M. Yurek and A. B. Tveten were instrumental in providing technical and logistical support for both the heading sensor test and the MARS prototype test. A. Dandridge (NRL) is acknowledged for helpful technical discussions and administrative support for all the NRL work summarized here.

Gunnar Wang, Svere Knudsen, Terje Lund, and Jan R. Nilssen of the Norwegian Defence Research Establishment were responsible for the electro-optical cable, all mechanical features of the sensor housing, and the sensor/cable interface for the MARS test.

REFERENCES

1. T. G. Giallorenzi, J. A. Bucaro, A. Dandridge, G. H. Sigel, Jr., J. H. Cole, S. C. Rashleigh, and R. G. Priest, "Optical fiber sensor technology," IEEE J. Quant. Electron., **QE-18**, 626-665 (1982).

2. F. Bucholtz, D.M. Dagenais and K.P. Koo, "High-frequency fibre-optic magnetometer with 70 fT/√Hz resolution, " Electron. Lett., **25**, 1719-1721 (1989).

3. D.M. Dagenais, F. Bucholtz, K.P. Koo, and A. Dandridge, "Detection of low-frequency magnetic signals in a magnetostrictive fiber-optic sensor with suppressed residual signal, " J. Lightwave Technol., **7**, 881-887 (1989).

4. F. Bucholtz, D. M. Dagenais, K. P. Koo and S. T. Vohra, "Recent developments in fiber optic magnetostrictive sensors," SPIE **Vol. 1367**, Fiber Optic and Laser Sensors VIII, 226 - 235 (1990).

5. F. Bucholtz, K. P. Koo, A. D. Kersey, and A. Dandridge, "Fiber optic magnetic sensor development," **SPIE Vol. 718**, Fiber Optic and Laser Sensors IV, 56 - 65 (1987).

6. G. L. Romani, S. J. Williamson, and L. Kaufman, "Biomagnetic instrumentation," Rev. Sci. Instrum., **53**, 1815 - 1845 (1982).

7. J. Lenz, "A review of magnetic sensors," Proc. IEEE, **78**, 973-989 (1990).

8. G.W. Day, "Compact fiber sensors for the measurement of low level electric currents," Technical Digest of the 4th Int. Conf. on Optical Fiber Sensors, OFS '86, 81-84 (1986).

9. F. Bucholtz, "Fiber optic magnetic sensors," Chap. 12 in <u>Fiber Optic Sensors: An Introduction for Engineers and Scientists</u>, John Wiley, 1991, (E. Udd, Editor).

10. H. B. Callen and T. A. Welton, "Irreversibility and generalized noise,"Phys. Rev., **83**, 34-40 (1951).

11. F. Bucholtz, J. E. Colliander, and A. Dandridge, "Thermal noise spectrum of a fiber-optic magnetostrictive transducer," Opt. Lett., **16**, 432 - 434 (1991).

12. D. M. Dagenais, F. Bucholtz, K. P. Koo, S. Vohra, and J. M. Pond, Detection of ferromagnetic resonance in metallic glass by fiber interferometric strain measurement," Appl. Phys. Lett. **58**, 546 - 547 (1991).

13. F. Bucholtz, A. D. Kersey, and A. Dandridge, "DC fibre-optic accelerometer with sub-μg sensitivity," Elextron. Lett., **22**, 451-453 (1986).

14. J.D. Livingston, "Magnetomechanical properties of amorphous metals," Phys. Stat. Sol. (A), **70**, 591-596 (1982).

15. H. Goldstein, <u>Classical Mechanics</u>, Addison-Wesley Publising Co., Reading, MA, 1950, pp. 107-109.

16. S. S. Patrick and F. Bucholtz, "Fiber optic heading sensor for the all-optical towed array (AOTA)," Submitted to J. Underwater Acous.

17. A. R. Davis, S. S. Patrick, A. Dandridge and F. Bucholtz, " Remote fibre-optic ac magnetometer," Electron. Lett., **28**, 271 - 273 (1992).

18. E. A. Nichols, H. F. Morrison, and J. Clarke, "Signals and noise in measurements of low-frequency geomagnetic fields," J. Geophys. Res., **93**, 13,743 - 13,754 (1988).

19. F. Bucholtz, J. A. McVicker, C. A. Villarruel, A. R. Davis, S. S. Patrick, K. G. Wathen, A. B. Tveten, and A. Dandridge, "Fiber optic low-frequency telemetry link for node motion studies: Preliminary report of the June 1991 MARS test," NRL Memorandum Report 6934, January 20, 1992.

20. K.P. Koo, F. Bucholtz and A. Dandridge, "Sideband noise of a large phase carrier in interferometric sensors, " Electron. Lett., **23,** 1062-1063 (1987).

21. R. D. Rempt and C. Ramon, "Detection of cardiac magnetic field with fiber optic magnetic gradiometer," Accepted for publication, Photonics Technol. Lett., Jan. 1993.

22. D. Y. Kim, H. J. Kong, and B. Y. Kim, "Fiber-optic dc magnetic field sensor with balanced detection," IEEE Photon. Technol. Lett., **4,** 945 - 948 (1992).

Fiber Optic Smart Structures

Eric Udd
McDonnell Douglas Electronic Systems Company
Santa Ana, California   92705

The revolutions in the fiber optic telecommunication and optoelectronic industries have enabled the development of fiber optic sensors that offer a series of advantages over conventional electrical sensors.   This development in combination with advances in composite material technology have opened up the new field of fiber optic smart structures that offers mechanical and structural engineers the possibility of incorporating fiber optic nervous systems into their designs.

1.0   INTRODUCTION

Buildings and bridges that can call up central maintenance depots and report on their status after an earthquake, storm or simply with the passage of time, aircraft that "know" and communicate if it is safe to take off and monitor and correct for structural changes in flight and artificial limbs that can feel, react and touch are all manifestations of the dreams of engineers and scientists working on the emerging field of smart structures [1-9].   These dreams require materials that are lighter weight, have superior strength and the ability to change shape, degree of stiffness, mechanical and electrical properties as required.   The new materials must be "smart" with the ability to sense environmental changes within or around the structure and have the ability to interpret and react to these changes.

The realization of these dreams has required a nervous system capable of sensing change in the material while being part of the structure itself.   Fiber optic sensor technology has enabled the implementation of this nervous system by (1) providing sensors that are small and rugged enough that they can be integrated and consolidated directly into materials, (2) enabling sensors to be multiplexed in substantial numbers along a single line allowing weight reduction and minimizing points of ingress and egress into parts, (3) providing electrical isolation and immunity to electromagnetic interference, and (4) supporting the high bandwidth necessary for large numbers of high performance sensors. With the complementary revolutions taking place in the optoelectronic and fiber optic telecommunication industry, the designer of the fiber optic nervous system is continually being offered higher performance components at lower cost.

Development of full systems require the coordination of many disciplines including experts in the fields of materials, structures, actuators, signal processing, sensors and systems. This paper provides an overview of how these disciplines interact to form the field of fiber optic smart structures. The paper concludes with a review of some of the emerging application areas.

2.0 ASPECTS OF FIBER OPTIC SMART STRUCTURES

There are four main aspects to fiber optic smart structures. The first is the use of optical fibers embedded or attached to materials to augment the manufacturing process by monitoring such parameters as temperature, pressure, strain, degree of cure, chemical content and viscosity. Once a part has been made the next aspect of fiber optic smart structures is to enhance the nondestructive evaluation of parties. This aspect allows the assessment of the integrity of parts prior to assembly for damage or defects that may have arisen during manufactures or handling. The parts can then be assembled and sensors integrated to form a health management system to assess overall structural integrity.

The highest level of integration involves combining actuator systems with a health monitoring system and signal processing to form control systems. These systems could be used to augment flight control or enable a building to "react" to an earthquake in a manner that minimizes damages.

3.0 IMPLEMENTATION AND CHALLENGES OF FIBER OPTIC SMART STRUCTURES

The basic functions of a fiber optic smart structure system are to (1) sense environmental conditions in or around the structure, (2) convey the information back to an optical and or electronic signal processor, and (3) perform an action as a result of the sensed information. Figure 1 shows a basic system in block diagram form. An environmental effect acts on and or around a structure which has embedded and or attached multiplexed fiber optic sensors. The information is then conveyed via a fiber optic link to an optical/electronic signal processor. After processing the information is conveyed to a control system that may be tasked to do damage, performance or health management functions. This system may also have the capability of taking corrective action via actuators that are directed to react to the environmental phenomenon.

The technologies associated with fiber optic smart structures and their interrelationship is illustrated by Figure 2. The first major issue is embedding the optical fibers into the structure [10-15] in such a way that structural integrity is not compromised and ensuring that the interface between the optical fiber and the surrounding material allows accurate measurement of the environmental effects of interest. There are a wide range of materials of interest that include carbon epoxy, thermoplastics and concrete on the low end of the processing temperature range to titanium and carbon-

carbon for very high temperature applications. For most low to moderate temperature applications silica based optical fibers are adequate with a melting temperature of about 1400°C. Generally the practical limit is somewhat lower than this since at about 1000°C the migration of dopants from the core region of the fiber becomes a serious problem. Most designs using silica based fiber have an upper operating range of about 700°C to allow for an adequate margin. For very high temperatures work has been done on using optical fibers based on sapphire [16-17] that have the potential to operate up to about 2000°C. While most current work has been done using silica based optical fibers because of their low cost and availability sapphire fibers can be expected to play a significant role in the higher temperature regimes where very few sensors can survive.

To successfully embed optical fibers into materials the coating is critical. One of the first roles of the coating is to protect the optical fiber from moisture. This is due to significant strength degradation that may occur in the fiber when microcracks on the surface of the fiber are penetrated by moisture causing the cracks to propagate.

Another important role is that the coating must form an appropriate interface between the optical fiber and the host material allowing accurate measurement of this environmental effect to be monitored. For organic composite materials such as carbon epoxy or thermoplastics it is particularly important that the fiber coating is chemically compatible with the resin of the host material. As an example to measure strain in a thermoplastic material a polyimide coating would be an appropriate choice while an epoxy acrylate would not. The polyimide coating consolidates with the thermoplastic and provides an excellent interface to the glass while the epoxy acrylate does not consolidate well with the thermoplastic allowing a potential reduction in the inherent strength of the part and it does not provide a rigid interface to the glass allowing slippage under strain conditions. For the case of embedding optical fibers in metals the critical parameter is to match the metallurgy of the coating to that of the metal it is being embedded into. Examples would include using an aluminum coated fiber for embedding into an aluminum part or graphite aluminum or gold coatings for placing fibers into titanium [18-19]. In the case of reinforced concrete most of the concerns to date have focused on survival of the optical fiber and protection from moisture. It is less clear what coatings are most appropriate for nonhomogeneous materials and further work is being done. Once an appropriate coating has been chosen the next step is to embed the fiber into the material itself. In the case of composite materials it is important to consider the orientation of the fiber in the part as well as the overall size of the optical fiber and its coating [11, 13, 14, 20, 21]. Generally running the optical fiber parallel to the strength member fibers of the composite material minimizes potential strength degradation of the part. The orientation of the optical fiber also becomes less critical as the size of the fiber is decreased.

Using these two design considerations in combination with an appropriate fiber coating it is possible to embed large numbers of optical fibers into a composite part without significant strength degradation. After the optical fibers are in place the next major issue is ingress and egress out of the part and connectors [11, 22]. To successfully implement a connection it is extremely important to make adequate provision for strains relief and protection of the fiber at the point of ingress/egress as this is its most vulnerable point.

After the details of appropriate coatings, fiber size and placement into the part have been worked out it becomes necessary to select a fiber sensor [23-25] that will be able

to measure the environmental effect with sufficient sensitivity, dynamic range and scaling to meet the performance requirements of the system designer.
Table 1 shows some of the fiber sensors that have been used to support fiber optic smart structure systems. The classic interferometric fiber sensors namely the Mach-Zehnder, Michelson and Sagnac interferometers have mainly been used to support characterization studies [26-28]. While these sensors have high sensitivity and can be made to have good dynamic range it is difficult to embed several of these sensors mutliplexed together due to the large number of fiber optic leads and components. The distributed fiber sensors [29-31] offer the prospect of supplying environmental information at locations along a single fiber. While the long term prospects of these sensors look promising the low sensitivity and spatial resolution have caused many researchers to look at alternative approaches. The two that are most promising in terms of multiplex potential, sensitivity and spatial resolution and that are the focus of a good deal of activity are the Fabry-Perot Etalon [32-39] and the fiber grating [40-43]. These sensors also have the advantages of allowing a single point of ingress and egress and have spectrally dependent signals that minimize problems associated with variable losses associated with connectors and fiber leads. Two relatively low cost options the microbending [44, 45] and break [9] fiber sensors are available for applications that have less stringent requirements on accuracy. Two other sensors that are more accurate but are difficult to multiplex are the dual mode [46-49] and polarimetric [50] fiber sensors. Most of the work done so far with these sensors has been for single sensor configurations.

After the sensors are selected the next issue facing the fiber optic smart structure designer is the choice of multiplexing techniques necessary to support the required number of sensors along a single fiber. There are five basic techniques; wavelength, time division, frequency, polarization and coherence multiplexing [29,51]. The two most commonly used techniques are wavelength division multiplexing where each sensor is color encoded and time division multiplexing that can be used to separate sensors out spatially. Figure 3 illustrates a simple example of combining time and wavelength division multiplexing. In this case a spectrally broadband light source is pulsed and the resultant light beam propagates down an optical fiber reaching the sensors 11, 12, and 13 which reflect light centered about the wavelengths λ_1, λ_2 and λ_3 that are within the spectral bandwidth of the source. The resultant sensor signals are reflected back toward the light source and are coupled via the fiber beamsplitter to the dispersive elements which separates the colors centered about λ_1, λ_2 and λ_3 onto the three output detectors 1, 2, and 3 respectively. Similarly a second set of signals results from reflections from the second set of fiber sensors 21, 22 and 23. The process can be repeated until optical loss budgets are exceeded with subsequent sets of sensors. It is also possible to use frequency and coherence multiplexing effectively. Polarization multiplexing has been used mainly for two sensors in line and has had limited utility. In general the type of multiplexing technique selected will depend strongly of the fiber sensor utilized and its expected performance characteristics.

The next issue to be aced involves the processing of the information from the multiplexed fiber sensors. Some of the processing may be handled optically via filters or active optical elements. Electronic signal processing would then be used to complete the transformation of the signals to a form usable by the control system. Since large numbers of sensors and corresponding data may be involved there is considerable interest in using neural networks [52-56] to process the data.

The control system is defined by a System Designer who in turn imposes performance specifications on the sensors, multiplexing techniques and processing. To make the fiber optic smart structure system work effectively all of these technologists must work together through the design and development process. Since many of these technologists come from disciplines that have traditionally had little interaction this is one of the great challenges in this field. At the same time it is also one of its strong points as so little has been done in the past there remains a good deal of progress to be made in the future.

4.0 APPLICATIONS OF FIBER OPTIC SMART STRUCTURES

One way to look at fiber optic smart structures is to use the analogy to living things. The basic idea is to develop a fiber optic nervous system capable of collecting information on the condition of the structure/organism. The information is then relayed back to a processor/brain where the information is sorted and analyzed. Based on this information actuators/muscles may be activated, processing/body temperature raised or lowered, or chemicals/hormones injected into the structure/organism. The contribution of fiber optics is primarily to allow the formation of nervous systems under conditions that were at best extremely difficult with prior art technology and on a scale that offers orders of magnitude more information with simultaneous size and weight savings due to the high bandwidth/short wavelength nature of light.

To illustrate this potential the rest of this section is devoted to a series of application areas that hold considerable promise for the near term realization of the potential of fiber optic smart structures.

4.1 Aerospace applications

As an example of the application of fiber optic smart structures to an aerospace application consider the case of a launch vehicle such as a rocket. This type of platform consists of a series of subsystems that could potentially benefit by the integration of fiber optic smart structure systems. The first usage of these systems will be in the manufacture of composite parts. In the case of a launch vehicle this could involve sensors that could be used to support the manufacture of cryogenic tanks, rocket nozzles, farings, solid rocket booster casings and interstages. Figure 4 illustrates the case for a tank that is being filament wound. Fiber sensors could be wound in directly with the preimpregnated material and used to monitor the consolidation process. These sensors could then be used to augment nondestructive evaluation techniques prior to installation on the rocket. Once the tank has been installed these fiber sensors or others that might be placed at a different stage of the manufacturing process would be used as part of a vehicle health monitoring system to monitoring structural changes of the tank and warn of leaks. Other potential application areas on launch vehicles would include monitoring rocket nozzles for areas of excessive burning, augmenting separation systems between stages, vibration and acoustic damping of the payload and other key components and eventually support of vehicle shape control systems that would allow adaptive guidance.

Space based applications for fiber optic smart structures can also be expected to include large space platforms. In this case there are future needs for space based platforms that are extremely light weight and yet rigid. A section of this type of structure is illustrated by Figure 5. Fiber optic smart structures for this type of platform can be expected to manifest themselves in the form of health monitoring systems that would assess changes in the structural integrity of the platform due to such events as docking, impacts or on orbit aging. It can also be expected that fiber optic smart structures would be used in combination with actuator systems for vibration control in order to isolate critical areas requiring a zero g environment and to damp out oscillations that would degrade pointing and tracking accuracy. In some cases these platforms could include habitats such as that shown in Figure 6. Here the potential exists for building fiber optic damage assessment systems directly into the walls of the structure to monitor the location and severity of an impact. This type of system could potentially be designed to support a self healing structure which detects and assesses damage and then automatically initiates repair. Other features that could be built into the habitat include distributed acoustic sensors to locate gas leaks and radiation, electromagnetic, pressure and temperature sensors used to measure the environment around the structure.

For the near future commercial and military aircraft can be expected to be the first major aerospace users of fiber optic smart structures. Initial application areas would include the manufacturing and processing of parts and the introduction of simple health monitoring and structural control systems. Examples of these early systems would be icing indicators, vibration and localized strain monitoring systems. Later efforts would include automatic maintenance systems that would perform preflight and postflight assessments of the aircraft. These systems would also be used to improve and simplify repair procedures and to provide in flight structural integrity warning systems. Finally fiber optic smart structures would evolve into an integral part of smart aircraft. This could feature deformable structures that radically change shapes as required, self healing systems performing damage assessment, flight corrections and repair in flight, and automatic flight control systems featuring environmental awareness around the aircraft and real time corrective actions.

The first fiber optic smart structure systems are likely to consist of a small number of sensors. To fully access the power of this technology on large platforms thousands and in some cases tens of thousands of sensors and or discernable sensing points will be required. As an example consider the case of a composite wing where it may be desirable to measure strain every ten centimeters in order to fully map out the strain field. For a modest twenty square meter wing this would involve 2000 sensors. Because of processing requirements it is likely that a large aerospace platform will have two classes of fiber sensors. The first would be distributed fiber sensors [29-31] that would be used to localize an event such as damage. The second set of fiber sensors would be discrete strings of fiber optic sensors that could be used to support detailed assessments. In order to be practical these sensors must meet a series of requirements. The first and most important is that the fiber sensors be low in cost and that the function performed by the fiber optic smart structure system they support must be of significant value to the end user in order to justify the increased cost. As examples the fiber optic smart structure system might be used to enhance maintainability, improve reliability, or enhance the performance of the platform. In each case these improvements have definable economic value and the fiber optic smart structure system must pass the test of improving the overall value of the platform to the end user. Additional customer driven considerations that are important are the redundancy and survivability of the system, installation of the system onto the platform and repairability. Requirements on the fiber optic sensors that are driven by structural considerations and manufacturing are that the fiber sensors have a single point of ingress/egress, the fiber sensors should be no larger or very close to the diameter of the fiber, and that a significant number of sensors must be multiplexed along a single fiber line. When existing fiber optic sensors are considered two candidates for meeting these requirements are fiber optic grating [40-43] and etalon based fiber sensors [32-39]. Potentially these sensors may be multiplexed into "strings" of fiber optic sensors. In order to hold costs down these strings could be multiplexed using an optical switch that could be used to interrogate sensors as needed. A block diagram of a portion of the fiber optic smart structure architecture is shown in Figure 7. A fiber sensor demodulator is used to extract the data from the sensors in the string being accessed. The data is then formatted and transmitted to a system signal processor which in turn transfers the data to the vehicle health management/damage assessment system. As an example of how this data would be handled on an aerospace platform Figure 8 illustrates an avionics system for a fighter aircraft in block diagram form. The data transmitted from the fiber optic sensor system signal processor would be transferred to the vehicle health management bus. This in turn would be transferred to the vehicle management system data processors and used to support the other elements of the avionics system.

4.2   Medical applications and biological analogies

Some smart structure medical products already exist. One employs sensors to determine the amount of medication in the blood and readjusts the rate at which medication is intravenously supplied. Another would be closed loop treadmills where a patient's exertion level is adjusted to correspond to his heart rate.

Possibilities however abound for future fiber optic smart structure based products including artificial limbs that readjust themselves as required by the users action, artificial hearts that include oxygen, pressure and temperature sensors to regulate heart beat, beds that sense the pressure distribution of patients lying on them and adjust themselves to avoid bedsores and rehabilitation systems that help the patient enough to successfully complete exercise but act to challenge the patient. Virtually any biological system is consequently a candidate for replacement by a fiber optic smart structure based system. For medical applications fiber optics hold the additional advantage of using passive dielectric devices that do not pose an electrical shock threat or radiation electromagnetic energy.

As an example of a biological analogy consider the example of the tree of Figure 9. Its leaves orient themselves toward sunlight by sensing light and changing hydraulic actuators. The tree's roots search out and grow toward water and the overall growth pattern of the tree adopts to wind loading. One can easily carry these analogies over when designing a building. The building can be aerodynamically shaped, the glass tint may be made to change as light intensity varies and pilings with actuators or rollers could be used to help the building to adjust to windloads or an earthquake.

4.3 Civil structure applications

In addition to integrating fiber optic smart structures into buildings, efforts are underway to place fiber optics smart structures into highways, bridges and dams [57-60]. Many of these systems are being placed to monitor the long term health of the structures but they will also be used to augment damping and for control systems such as internal climate control for buildings.

One can easily go further by linking systems together between buildings that could serve to provide communities with fire protection, and emergency services as well as monitor such functions as gas and electric usage.

5.0 SUMMARY

Fiber optic smart structures are an enabling technology that will allow engineers to add nervous systems to their designs enabling damage assessment, vibration damping and many other capabilities to structures that would be very difficult to achieve by other means. This technology can be expected to have dramatic future impact in the aerospace, medical and civil structure fields.

REFERENCES

1. E. Udd, Editor, "Fiber Optic Smart Structures and Skins," Proceedings of SPIE, Vol. 986, Boston, September 1988.
2. E. Udd, Editor, "Fiber Optic Smart Structures and Skins II," Proceedings of SPIE, Vol. 1170, Boston, September 1989.
3. E. Udd and R. O. Claus, Editors, "Fiber Optic Smart Structures Skins III," Proceedings of SPIE, Vol. 1370, San Jose, September 990.
4. R. O. Claus and E. Udd, Editors, "Fiber Optic Smart Structures and Skins IV," Proceedings of SPIE, Vol. 1588, Boston, September, 1991.
5. G. J. Knowles, Editor, "Active Materials and Adaptive Structures," Proceedings of the ADPA/AIAA/ASME/SPIE Conference on Active Materials and Adaptive Structures, IOP Publishing, November, 1991.

6. B. Culshaw, P. T. Gardiner and A. McDonach, Editors, "First European Conference on Smart Structures and Materials," Proceedings of SPIE, Vol. 1777, May, 1992.

7. R. O. Claus and R. S. Rogowski, Editors, "Fiber Optic Smart Structures and Skins V," Proceedings of SPIE, Vol. 1798, September, 1992.

8. E. Udd, "Embedded Sensors Make Structures "Smart," Laser Focus, p. 138, May 1988.

9. R. M. Measures, "Smart Structures in Nerves of Glass," Progress in Aerospace Science, Vol. 26, p. 289, 1989.

10. E. H. Urruti, P. E. Blaszyk, and R. M. Hawk, "Optical Fibers for Structural Sensing Applications," Proceedings of SPIE, Vol. 986, p. 158, 1988.

11. R. L. Wood, A. K. Tay, and D. A. Wilson, "Design and Fabrication Considerations for Composite Structures with Embedded Fiber Optic Sensors," Proceedings of SPIE, Vol. 1170, p. 160, 1989.

12. W. Maslach, Jr. and J. S. Sirkis, "Strain or Stress Component Separation in Surface Mounted Interferometric Optical Fiber Strain Sensors," Proceedings of SPIE, Vol. 1170, p. 452, 1989.

13. D. W. Jensen and J. Pascual, "Degradation of Graphite/Bismaleimide Laminates with Multiple Embedded Fiber Optic Sensors," Proceedings of SPIE, Vol. 1370, p. 228, 1990.

14. J. S. Sirkis and A. Dasgupta, "Optimal Coatings for Intelligent Fiber Optic Sensors," Proceedings of SPIE, Vol. 1370, p. 129, 1990.

15. C. DiFrancia, R. O. Claus and T. C. Ward, "Role of Adhesion in Optical-Fiber-Based Smart Composite Structures and its Implementation in Strain Analysis for Modeling of an Embedded Optical Fiber," Proceedings of SPIE, Vol. 1588, p. 44, 1991.

16. A. R. Raheem-Kizchery, S. B. Desu and R. O. Claus, "High-Temperature Refractory Coating Materials for Sapphire Waveguides," Proceedings of SPIE, Vol. 1170, p. 513, 1989.

17. K. A. Murphy, B. R. Fogg, C. Z. Wang, A. M. Vengsarker and R. C. Clans, "Sapphire Fiber Interferometer for Microdisplacement Measurements at High Temperatures," Proceedings of SPIE, Vol. 1588, p. 117, 1991.

18. S. E. Baldini, E. Nowakowski, H. G. Smith, E. J. Freible, M. A. Putnam, R. Royowski, L. D. Melvin, R. O. Claus, T. Tran, M. S. Helben, Jr., "Cooperative Implementation of a High-Temperature Acoustic Sensor," Proceedings of SPIE, Vol. 1588, p. 125, 1991.

19. S. E. Baldini, D. J. Tubbs, and W. A. Stange, "Embedding Fiber Optic Sensors in Titanium Matrix Composites," Proceedings of SPIE, Vol. 1370, p. 162, 1990.

20. A. Dasgupta, Y. Wan, J. S. Sirkis and H. Singh, "Micromechanical Investigation of an Optical Fiber Embedded in a Luminated Composite," Proceedings of SPIE, Vol. 1370, p. 129, 1990.

21. A. M. Vengsarker, K. A. Murphy, M. F. Gunther, A. J. Plante, and R. O. Claus, "Low Profile Fibers for Embedded Smart Structure Applications," Proceedings of SPIE, Vol. 1588, p. 2, 1991.

22. W. B. Spillman, Jr., "Fiber Optic Sensors for Composite Monitoring," Proceedings of SPIE, Vol. 986, p. 6, 1988.

23. E. Udd, Editor, "Fiber Optic Sensors: An Introduction for Engineers and Scientists," Wiley, New York, 1991.

24. J. Dakin and B. Culshaw, Editors, "Optical Fiber Sensors: Principles and Components," Vol. 1 Artech House, Boston, 1988.

25. B. Culshaw and J. Dakin, "Optical Fiber Sensors: Systems and Applications," Vol. 2, Artech House, Norwood, Mass, 1989.

26. D. A. Brown, B. Tan and S. L. Garret, "Nondestructive Dynamic Complex Moduli Measurements Using a Michelson Interferometer and a Resonant Bar Technique," Proceedings of SPIE, Vol. 1370, p. 238, 1990.

27. J. A. Sirkis and H. W. Haslach, Jr., "Complete Phase-Strain Model for Structurally Embedded Interferometric Optical Fiber Sensors," Proceedings of SPIE, Vol. 1370, p. 248, 1990.

28. E. Udd, R. J. Michal, S. E. Higley, J. P. Theriault, P. LeCong, D. A. Jolin and A. M. Markus, "Fiber Optic Sensor Systems for Aerospace Applications," Proceedings of SPIE, Vol. 838, p. 162, 1987.

29. A. D. Kersey, "Distributed and Multiplexed Fiber Optic Sensors," in "Fiber Optic Sensors: An Introduction for Engineers and Scientists," edited by E. Udd, Wiley, 1991.

30. J. P. Dakin, D. A. J. Pearce, A. P. Strong, and C. A. Wade, "A Novel Distributed Optical Fiber Sensing System Enabling Location of Disturbances in a Sagnac Loop Interferometer, Proceedings of SPIE, Vol. 838, p. 325, 1987.

31. E. Udd, "Sagnac Distributed Sensor Concepts," Proceedings of SPIE, Vol. 1586, p. 46, 1991.

32. C. E. Lee and H. F. Taylor, "Interferometric Optical Fiber Sensors Using Internal Mirrors," Electronic Letters, Vol. 24, p. 193, 1988.

33. C. E. Lee, R. A. Atkins and H. F. Taylor, "Performance of a Fiber-Optic Temperature Sensor from -200 to 1050°C," Optics Letters, Vol. 13, p. 1038, 1988.

34. T. Valis, D. Hogg and R. M. Measures, "Fiber Optic Fabry-Perot Strain Sensor," IEEE Photonics Technology Letters, Vol. 2, p. 227, 1990.

35. C. E. Lee and H. F. Taylor, "Fiber Optic Fabry Perot Sensor Using a Low Coherence Source," IEEE Journal of Lightwave Technology, Vol. 9, p. 129, 1991.

36. C. E. Lee, H. F. Taylor, A. M. Markus and E. Udd, "Optical Fiber Fabry-Perot Embedded Sensor," Optics Letters, Vol. 14, p. 1225, 1989.

37. T. Valis, D. Hogg, and R. M. Measures, "Composite Material Embedded Fiber Optic Fabry-Perot Strain Rosette," Proceedings of SPIE, Vol. 1370, p. 154, 1990.

38. K. A. Murphy, B. R. Fogg, G. Z. Wang, A. M. Vengsarkar and R. O. Claus, "Sapphire Fiber Interferometer for Microdisplacement Measurements at High Temperatures," Proceedings of SPIE, Vol. 1588, p. 117, 1991.

39. K. A. Murphy, M. F. Gunther, A. M. Vengsarkar and R. O. Claus, "Fabry Perot Fiber Optics Sensors in Full Scale Testing on an F- 15 Aircraft," Proceedings of SPIE, Vol. 1588, p. 134, 1991.

40. H. D. Simonsen, R. Paetsch, and J. R. Dunphy, "Fiber Bragg Grating Sensor Demonstration in Glass Fiber Reinforced Polyester Composite," Proceeding of First European Conference on Smart Structures and Materials, p. 73, Glasgow, 1992.

41. G. Meltz, W. W. Morrey, and W. H. Glenn, "Formation of Bragg Grating in Optical Fibers by a Transverse Holographic Method," Optics Letters, Vol. 14, p. 823, 1989.

42. W. W. Morey, "Distributed Fiber Grating Sensors," Proceedings of the 7th Optical Fiber Sensors Conference," p. 285, Sydney, 1990.

43. S. M. Melle, K. Liu, and R. M. Measures, "Strain Sensing Using a Fiber Optics Bragg Grating," Proceedings of SPIE, Vol. 1588, p. 255, 1991.

44. H. Smith, Jr., A. Garrett and C. R. Saff, "Smart Structure Concept Study," Proceedings of SPIE, Vol. 1170, p. 224, 1989.

45. E. Udd, J. P. Theriault, A. Markus, and Y. Bar-Cohen, "Microbending Fiber Optic Sensors for Smart Structures," Proceedings of SPIE, Vol. 1170, p. 478, 1989.

46. Z. J. Lu and F. A. Blaha, "A Fiber Optic Strain and Impact Sensor System for Composite Materials," Proceedings of SPIE, Vol. 1170, p. 239, 1989.

47. D. A. Cox, D. Thomas, K. Reichard, D. Lindner and R. O. Claus,"Model Domain Fiber Optic Sensor for Closed Loop Vibration Control of a Flexible Beam," Proceedings of SPIE, Vol. 1170, p. 372, 1989.

48. B. Y. Kim, J. N. Blake, S. Y. Huang and H. J. Shaw, "Use of Highly Elliptical Core Fibers for Two-Mode Fiber Devices," Optics Letters, Vol. 12, p. 729, 1987.

49. J. N. Blake, S. Y. Huang, B. Y. Kim and H. J. Shaw, "Strain Effects on Highly Elliptical Core Two Mode Fibers," Optics Letters, Vol. 12, p. 732, 1987.

50. W. B. Spillman, Jr., L. B. Maurice, J. R. Lord, and D. M. Crowne, "Quasi-distributed Polarimetric Strain Sensor for Smart Skins Applications," Proceedings of SPIE, Vol. 1170, p. 483, 1989.

51. A. D. Kersey and J. P. Dakin, Editors, "Distributed and Multiplexed Fiber Optic Sensors," Proceedings of SPIE, Vol. 1586, 1991.

52. B. G. Grossman, H. Hou, R. H. Nassar, A. Ren, and M. H. Thursby, "Neural Network Processing of Fiber Optic Sensors and Arrays," Proceedings of SPIE, Vol. 1370, p. 205, 1990.

53. M. Thursby, K. Yoo, and B. Grossman, "Neural Control of Smart Electromagnetic Structures," Proceedings of SPIE, Vol. 1588, p. 219, 1991.

54. B. Grossman, X. Gao and M. Thursby, "Composite Damage Assessment Employing an Optical Neural Network Processor and an Embedded Fiber Optic Sensor Array," Proceedings of SPIE, Vol. 1588, p. 64, 1991.

55. J. M. Mazzu, S. M. Allen and A. K. Caglayen, "Neural Network/Knowledge Based Systems for Smart Structures," Proceeding of the Active Materials and Adaptive Structures Conference, p. 243, Alexandria, VA, 1991.

56. M. R. Napolitano, C. I. Chen and R. Nutter, "Application of a Neural Network to the Active Control of Structural Vibration," Proceedings of the Active Materials and Adaptive Structures Conference, p. 247, Alexandria, VA, 1991.

57. D. R. Huston, "Smart Civil Structures-An Overview," Proceedings of SPIE, Vol. 1588, p. 182, 1991.

58. D. R. Huston, P. L. Fuhr, P. J. Kajenski, T. P. Ambrose, and W. B. Spillman, "Installation and Preliminary Results from Fiber Optics Sensors Embedded in a Concrete Building," Proceedings of the First European Conference on Smart Structures and Materials, p. 409, Glasgow, 1992.

59. H. D. Wright and R. M. Lloyd, "Monitoring the Performance of Real Building Structures," Proceedings of the First European Conference on Smart Structures and Materials, p. 219, Glasgow, 1992.

60. A. Holst and R. Lessing, "Fiber-Optic Intensity-Modulated Sensors for Continuous Observation of Concrete and Rock-Filled Dams," Proceedings of the First European Conference on Smart Structures and Materials, p. 223, Glasgow, 1992.

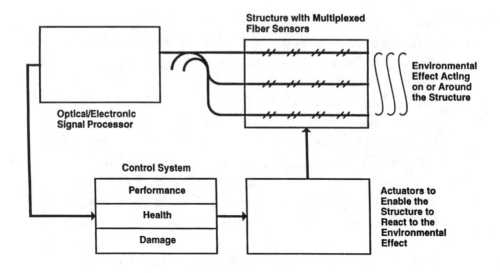

Figure 1.   Basic block diagram of a fiber optic
           smart structure system.

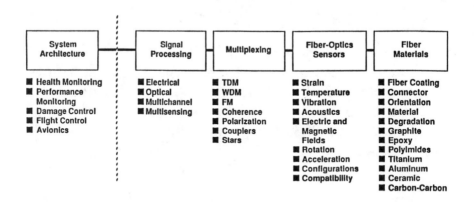

Figure 2.   Technologies associated with
           fiber optic smart structures.

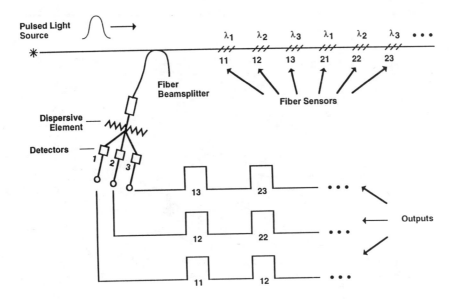

Figure 3.   Using time and wavelength division multiplexing to support several fiber sensors along a single fiber.

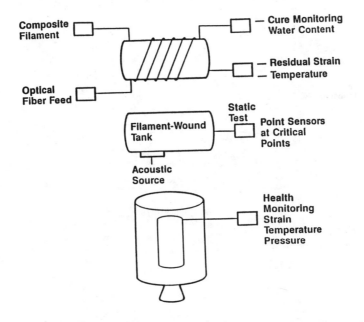

Figure 4.   Usage of fiber sensors to support the manufacture of a cryogenic tank and a health monitoring system.

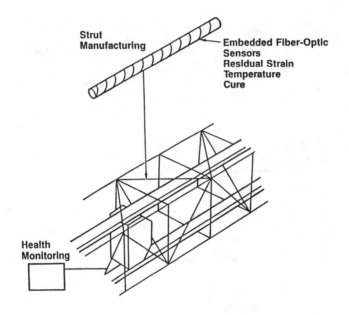

Figure 5.   Fiber optic smart structures used to
support a space based platform.

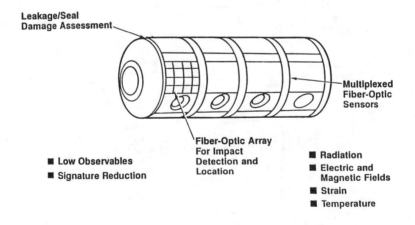

Figure 6.   Fiber optic damage and health
monitoring systems for a space habitat.

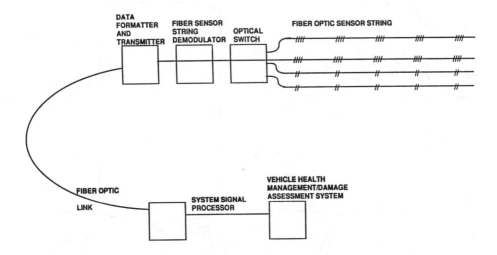

Figure 7.   Partial system supporting a fiber optic smart structure architecture.

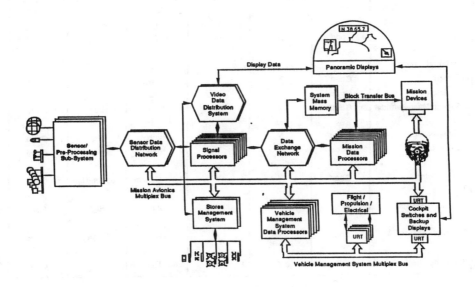

Figure 8.   Integrated avionics system using the Air Force pave pillar structure.

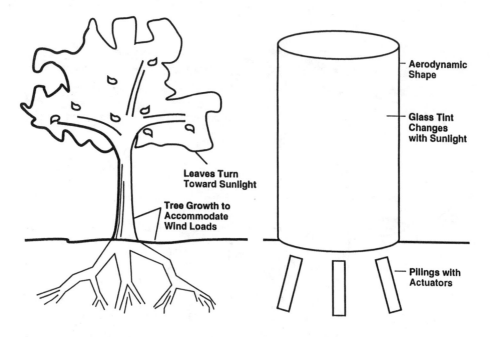

Figure 9.   Biological analogies to smart structures.

	Multiplexing Capability	Single Fiber Configurations	Single Ended Configuration	Sensitivity	Suitability for Short Range Length	Embedibility 10 Sensing Locations
❑ Sagnac/Mach-Zehnder	Good	N	Y	High	Low	Poor
❑ Michelson	Good	N	Y	High	Medium	Poor
❑ Sagnac	Good	N	N	Medium	Low	Poor
Distributed Sensors						
❑ Raman	Good	Y	Y	Low	Low	Good
❑ Rayleigh	Good	Y	Y	Low	Low	Good
❑ Sagnac/Mach-Zehnder	Good	Y	N	Low	Low	Good
❑ Fabry-Perot Etalon	Good	Y	Y	High	High	Good
❑ Fiber Grating	Good	Y	Y	High	High	Good
Microbending	Good	Y	Y	Low-Medium	Medium-High	Good
Dual Mode Sensor	Poor	Y	Y	High	High	Poor
Polarization	Poor	Y	Y	High	High	Poor-Medium
Break Sensors	No	Y	Y	Low- Medium	High	Good

Table 1.   Representative fiber optic sensor candidates for smart structures

Status and review of fiber optic sensors in industry

John W. Berthold III

Babcock & Wilcox — Research and Development Division
1562 Beeson Street, Alliance, Ohio 44601

ABSTRACT

Progress continues in the development of fiber optic sensors in industry. Most applications still appear to be in special niche areas, and characterized by a relatively small number of units sold. Signs are present, however, that indicate this situation may be changing. This presentation will include specific examples of recent development work and fiber optic sensor hardware used in applications such as manufacturing, power generation, and chemical processing. An assessment will be made of sensor performance in present applications, and progress made over the last several years in matching the technology and capability of fiber optic sensors to industrial needs. Both the shortcomings and advantages of fiber optic sensors will be discussed along with the outlook for future new applications, the direction fiber sensor technology is going, and why.

1. INTRODUCTION

Specialty fiber optic sensors continue to be developed for industrial applications. Significant progress has been made since I last reviewed the subject of fiber optic sensors in industry (see *Proc. SPIE 838*, p. 2 (1987)). Test and demonstration projects have been conducted over the last several years, and several of these efforts are summarized in this paper. Applications are niche-driven and tend to be in those areas where fiber optics offers distinct advantages. Examples include pressure measurements in 400°C reactors and temperature measurements in high voltage transformers. However, the sensor industry as a whole is primarily driven by special applications for specialty sensors. Little universality is present in this industry. Thus, fiber optic sensors are simply one more technology to find a home among the potpourri of sensor and transducer technologies, each of which is engineered into special products for special customer needs.

Of the many individual fiber sensor types (generically categorized as interferometric, polarization, wavelength, and intensity) demonstrated in the lab, fiber intensity sensors are most advanced from an engineering and commercial standpoint. Optical and fiber optic intensity sensors predominate in industrial applications for several reasons. These include:

- simple, manufacturable sensor assemblies
- simple, low-cost signal processors
- available self-referencing methods to significantly reduce link sensitivity
- compatibility with optical time domain reflectometry methods for sensor multiplexing

These factors favor intensity sensors for near-term commercial industrial applications. However, other sensor types, especially in-fiber Fabry-Perot and in-fiber grating sensors show promise in the longer term.

In this paper, examples are provided of some specific improvements in fiber sensor technology, and the widening areas of opportunity for fiber sensor industrial applications are discussed.

2. TECHNOLOGY IMPROVEMENTS

Recent improvements in fiber optic intensity sensor technology have resulted in performance competitive with existing commercial sensor technologies.

2.1 Pressure measurement

One example is the microbend fiber optic pressure transducer, whose performance has evolved along with the engineering of the sensor head.

Figures 1, 2, 3, and 4 illustrate how the accuracy of a microbend fiber optic pressure transducer has steadily improved over the last eight years[1,2,3]. Although these three calibration curves represent results for devices built for different pressure ranges, each pressure transducer contained a microbend fiber optic sensor. The data in Figure 1 obtained eight years ago exhibits a hysteresis loop. The data shown in Figure 2, obtained six years ago, is free from hysteresis but exhibits nonlinearity. The data shown in Figure 3, obtained six months ago, is practically free from the hysteresis and nonlinearity. The measurement of nonlinearity from Figure 3 is plotted in Figure 4 and amounts to less than ±1%. Overall non-repeatability and hysteresis obtained from multiple calibrations is less than ±0.1%.

The pressure transducer head that produced the data in Figures 3 and 4 is shown in Figure 5. This advanced prototype package was designed for manufacturability. The transducer head is a totally sealed assembly where the final seal welds were made using an automated electron beam welding process. This transducer is capable of measuring absolute pressure (vacuum reference).

The transducer consists of a diaphragm which deflects in proportion to applied pressure. The deflecting diaphragm directly modulates light transmitted through a microbend fiber optic intensity sensor[2], where a multimode step-index fiber is squeezed between a pair of corrugated tooth blocks. One tooth block is positioned at the center of the diaphragm and moves as the diaphragm deflects. The other tooth block remains stationary. Diaphragm deflection causes a change in the amplitude of the periodic distortion of the fiber squeezed between the tooth blocks, and this mechanical change in amplitude results, in turn, in a proportional change in optical power transmitted through the fiber. Thus, a measurement of the change in transmitted power provides an analog signal proportional to applied pressure. The calibration data (Figure 3) was obtained with the transducer (Figure 5) assembled in a hardened enclosure along with a fiber optic temperature reference, multimode wavelength division demultiplexer, and fused bi-directional coupler.

2.2 Vibration measurement

As an important part of sensor performance improvements, efforts are also underway to lower the cost of fiber optic intensity sensors. Work is now underway on the first contract order aimed at low-cost production of volume quantities of microbend fiber optic accelerometers. This work is being performed by Optical Sensors and Systems, Newark, NJ, for the

Naval Weapons Support Center, Crane, IN. Microbend accelerometers contain a microbend fiber optic intensity sensor that functions similarly to the one described above in the pressure transducer. Performance of microbend vibration sensors and accelerometers has been reported in previous demonstration projects conducted by Optical Technologies, Herndon, VA[4], and Babcock & Wilcox, Alliance, OH[5].

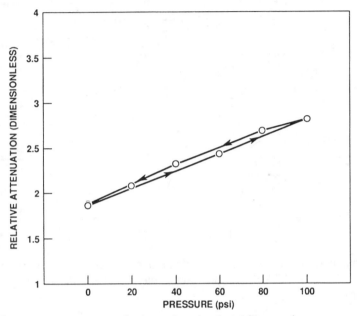

Fig. 1. Output versus pressure for an early microbend fiber optic pressure transducer built for NASA. Note presence of hysteresis.

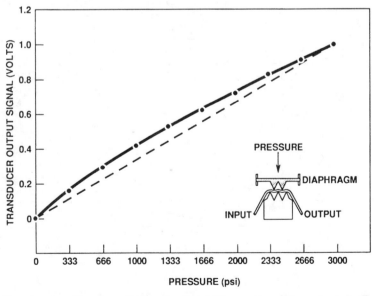

Fig. 2. Output versus pressure for a microbend fiber optic pressure transducer built for DOE. Note presence of nonlinearity.

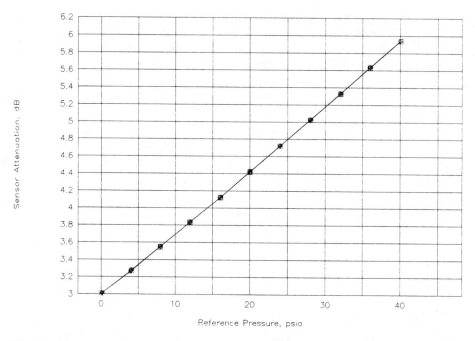

Fig. 3. Output versus pressure for a recent microbend fiber optic pressure transducer built for McAir and NASA FOCSI program. Note virtual elimination of hysteresis and nonlinearity.

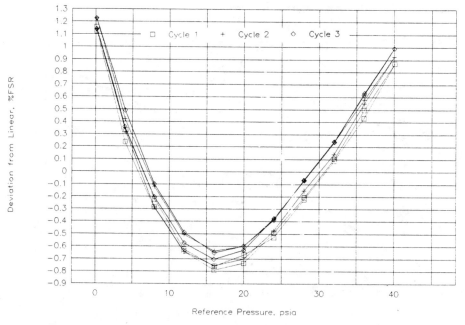

Fig. 4. Deviation from linearity versus pressure and repeatability for a microbend fiber optic pressure transducer built for McAir and NASA FOCSI program.

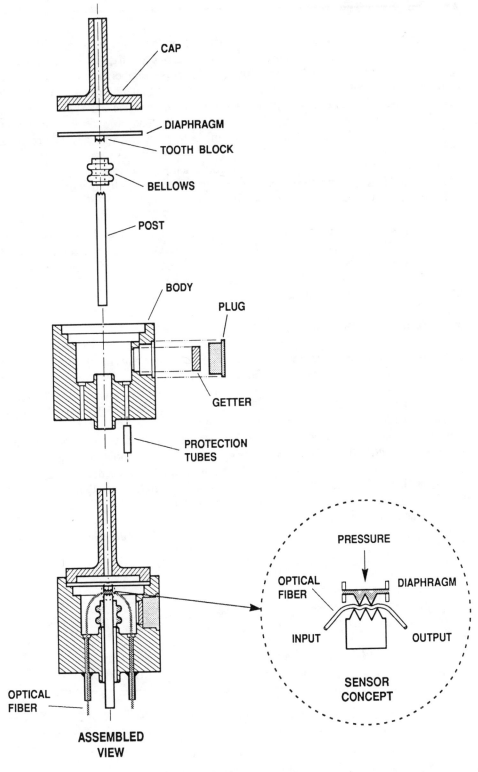

Fig. 5. Cross-section diagram of a microbend fiber optic pressure transducer.

2.3 Temperature measurement

Temperature sensor applications have been identified in the utility industry, where fiber optics has a distinct advantage over tried and true thermocouples. In these applications, EMI resistance is the primary requirement. Distributed fiber optic temperature sensing has been performed in high voltage circuit breakers by workers from Battelle Columbus Labs. This work was done at Ohio Edison's Burger plant by Boiarski and Nilsson[6]. They used an optical time domain reflectometer (OTDR) distributed sensing method and special sensing fiber, with high refractive index cladding material made by both 3M Specialty Fibers, West Haven, CT, and Fiberguide Industries, Stirling, NJ. System performance achieved in this application was $\pm2°C$ accuracy over a $0°C$ to $150°C$ temperature range. Sensor plus link lengths were a total of about 30 m. The system used an OTDR with 10 cm length resolution and 100/140 multimode transmission link fiber.

In a somewhat related application, Luxtron, Mountain View, CA, has been working with Pacific Gas & Electric to develop temperature measurement methods for detecting hot spots in the windings of power plant transformers[7]. Luxtron has successfully applied its well-known fluorescent decay time fiber optic sensor in this application to monitor high voltage transformer winding temperature.

2.4 Multimode sensor multiplexing

Although most applications and demonstrations up to this time have involved single rather than multiple sensors, it has become clear that the multiplexing capability of fiber optics is an important and potentially low-cost advantage of this technology. For example, multiple fiber optic accelerometers can be linked to a common transmission "bus" fiber. These accelerometers can all be remotely interrogated in turn from a central processor without any intervening electronics. Either wavelength division[8] or time division[9] multiplexing methods could be employed using demonstrated methods that exist today. In contrast, commercially available piezo-electric accelerometers each requires its own charge amplifier/ signal conditioner to achieve low noise operation, and thus these devices are not compatible with a low-cost multiplexing architecture. The differences in both hardware and complexity between fiber optic and piezoelectric accelerometers are readily apparent from examination of the two architectures shown in Figure 6 and 7.

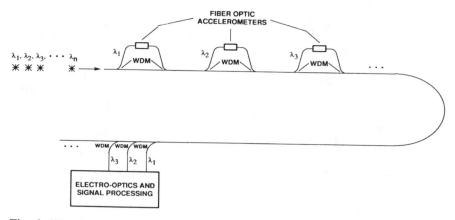

Fig. 6. Wavelength division multiplexed fiber optic intensity sensor accelerometers.

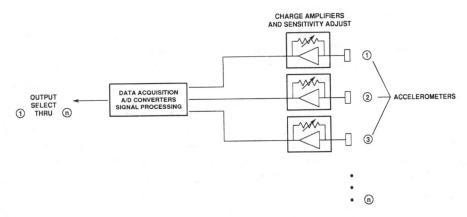

Fig. 7. Electronically multiplexed piezoelectric accelerometers.

2.5 Analog intensity sensor self-referencing

Self-referencing methods needed to reduce or eliminate link sensitivity of fiber intensity sensors have evolved into two effective methods, one based on optical wavelength normalization[10] and the other on time division normalization[11]. Both methods have been shown to reduce the effects of link attenuation and connector losses to <0.25%. Further improvements are possible. These methods effectively make fiber sensors smart without the need for local electronics processing at the sensor location. Experience at Babcock & Wilcox with both wavelength and time division normalization methods is described in this section.

One approach for wavelength division normalization to self-reference a fiber optic pressure transducer is shown in Figure 8. This approach was implemented with the transducer, described in Section 2.1, to produce the data plotted in Figures 3 and 4. Optical power is provided over a broad wavelength band at the sensor input. A wavelength demultiplexer (WDM) divides the input into two separate wavelength channels centered at λ_1 and λ_2. The

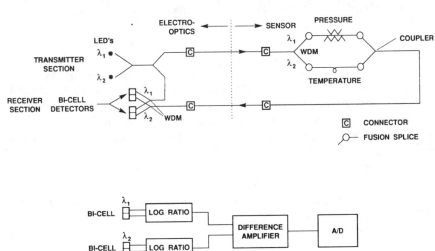

Fig. 8. Method for wavelength division normalization and signal processing.

channel centered at λ_1 is intensity modulated by the microbend pressure sensor and the channel centered at λ_2 acts as an intensity and/or temperature reference. Light intensity in both channels is modulated proportionally by link changes such as cable bends and connector losses. At the microbend pressure sensor output, the two wavelength channels are recombined and the light signals return to the electro-optic receiver, where they are again wavelength demultiplexed. The pressure signal is extracted from the λ_1 channel as shown in the lower half of Figure 8. First, a reference intensity light signal from each wavelength source in the electro-optic transmitter is used to compensate the corresponding signal channel from the sensor. The compensation may be performed using bi-cell photodetectors and log ratio current amplifiers. This operation on each sensor wavelength channel removes optical signal variations caused by light source intensity fluctuations. Next, the resultant signal voltages are subtracted in a difference amplifier. This operation removes optical signal variations caused by cable and connector losses, which affect both wavelength channels the same (to first order). Ideally then, all resulting changes in the difference amplifier output are proportional only to pressure induced changes in the microbend sensor in the pressure transducer head.

In practice, several precautions must be taken to make sure that performance as close as possible to the ideal described above is achieved. First, the WDMs in the sensor and receiver must be closely matched so that the spectra of the separated wavelength channels match. Otherwise, normalization performed by the log ratio operations will be incomplete. Second, the interchannel cross-talk in these WDMs must be minimized. Our analysis indicates that -30 dB cross-talk is more than adequate for this application. Third, temperature dependent changes in the channel spectra must be minimized. This requirement can be met using a grating based, rather than a dichroic WDM. Finally, the temperature sensitivity and repeatability of temperature characteristics of both the WDM and coupler are very important.

Time division intensity normalization (TDIN) is performed as diagrammed in Figure 9. This approach is based on methods developed at Eldec, Bothell, WA[11]. The fiber optic circuit in Figure 9 is straightforward to implement with a minimum number of components. It contains a single light source and detector, 2x2 fiber optic coupler, delay coil, microbend sensor, and silver coated fiber end reflectors.

Light from a pulsed source is transmitted to the pressure sensor head containing microbend fiber optic intensity sensor, delay coil, and 2x2 power splitter. Each input light pulse is divided at the splitter between the sensor tap and the delay coil tap. The ends of the taps are mirrored. If the round trip time through the delay coil is long enough, the received pulses from the sensor I_S and delay coil I_R will be separated in time (see timing diagram in Figure 9). Note that any link changes such as cable bending or connector mating and demating will introduce offsets which will affect the received pulses similarly. Thus, a ratiometric measurement I_S/I_R will provide an output signal free of any cable or connector dependent offsets.

A big advantage of this approach is the symmetric manner in which the coupler operates. Table 1 illustrates how insensitive the ratiometric measurement of I_S/I_R is to changes in the coupling ratio.

The total round trip optical delay length to match pulse arrival time to the TDIN receiver is about 13 meters at 1300 nm wavelength. This delay requires a coil length of about 6.5 meters. With 13 meter delay, the spacing between pulse centers is about 62.5 ns.

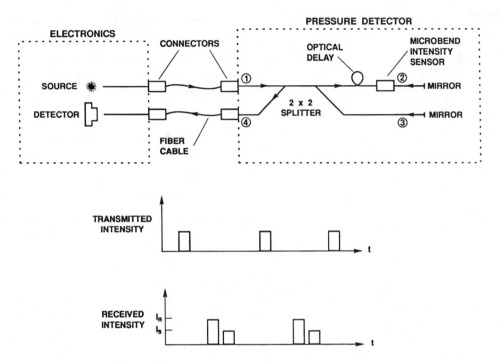

Fig. 9. Method for time division intensity normalization.

Table 1
Sensitivity to coupling ratio changes in symmetric 2x2 coupler with TDIN
as shown in Figure 9

Coupling Ratio	Input Intensity Port 1	Sensor Intensity Port 2	Reference Intensity Port 3	Output Intensities Port 4		
				I_S	I_R	I_S/I_R
.5/.5	1	.5	.5	.25	.25	1
.6/.4	1	.6	.4	.24	.24	1
.4/.6	1	.4	.6	.24	.24	1
.7/.3	1	.7	.3	.21	.21	1
.3/.7	1	.3	.7	.21	.21	1

3. APPLICATIONS AND NEW DEVELOPMENTS

Activities in various application areas are summarized in this section. The activities and applications discussed are not meant to be exhaustive, but rather representative, and to indicate the widespread application of fiber optic intensity sensors.

3.1 Aerospace

Work is underway on two phases of the NASA-sponsored Fiber Optic Control System Integration (FOCSI) project. In one phase (engines), General Electric is the prime and in the other phase (airframe) McDonnell Douglas (McAir) is the prime. The goal of the McAir effort is to flight test fiber optic instrumentation in parallel with existing instrumentation. A wavelength division multiplexed, electro-optic unit and sensors for temperature (time rate of fluorescence decay), pressure (analog microbend intensity), linear and rotary position (analog and digital code plate) have been delivered to McAir and are currently under test in the lab.

Activities in smart structures and skins for aerospace applications are discussed separately in these proceedings by E. Udd.

One indicator of increasing interest in aerospace applications of fiber sensors has come through an increasing number of inquiries from both the air frame manufacturers and their suppliers. These inquiries request detailed information on fiber sensor performance specifications and unit prices.

3.2 Automotive

Many automotive applications of fiber sensors are being explored. One engine application is the in-cylinder measurement of pressure for control of spark advance to eliminate engine knock. Several parallel efforts are underway at different organizations (Texas A&M University, Sandia, and NASA Langley) to develop miniature fiber optic pressure sensors which could be used for this application.

One major automobile manufacturer (General Motors) now installs plastic optical fiber as a GO/NOGO sensor to view headlights and other light sources throughout the vehicle. One end of each fiber is in the reflector compartment or adjacent to a light bulb. The other end of the fiber terminates in a visual display visible to the driver. A failed bulb in the lighting system appears as a dark spot in the visual display.

Related efforts are underway to evaluate plastic fiber for vehicle data networks. These efforts could ultimately lead to packaging of fibers with electrical lines in the vehicle wiring harness. Emphasis is placed on plastic fiber because it allows relatively short cable runs, low cost, and ease of termination at splices and connectors.

3.3 Environmental sensors and chemical process sensors

Many new fiber optic chemical sensors are being developed and are reviewed in these proceedings by R. A. Lieberman.

A high level of activity is also presently underway to support new U.S. Department of Energy programs for clean-up of hazardous, radioactive, and mixed wastes at several of the U.S. nuclear weapons production sites. Emphasis is on development and application of remote in-situ sensors for chlorinated hydrocarbons, industrial organic solvents, heavy metals, low level radioactive, and transuranic wastes. For the chlorinated hydrocarbons and some solvents, remote colorimetric optrodes have been developed and applied by Lawrence Livermore National Laboratory[12]. Sensitivities <10 ppb have been achieved. Fiber optic probes to enable in-situ ultraviolet absorption measurement for ground water monitoring have been demonstrated by workers at Oak Ridge National Laboratory[13]. This group is also working with EIC Laboratories, Norwood, MA, to develop surface-enhanced Raman sensors for in-situ detection of chlorinated hydrocarbons. At Babcock & Wilcox, a new project is underway to develop a scintillating fiber optic sensor for post-closure monitoring of buried radioactive waste.

In the chemical process industry, Fourier transform infrared spectroscopy (FTIR) performed remotely through infrared transmitting fiber has been shown to be useful in verification of proper epoxy cure during manufacture of composite materials[14].

Monitoring the transport and deposition of corrosion products in heat transfer and other processes fluids has been shown to be feasible using remote Raman spectroscopy through optical fibers[15]. Commercial hardware for remote Raman spectroscopy is available from several sources, one of which is Chromex, Albuquerque, NM. Monitoring localized corrosion build-up using fiber optic strain gages[16] is a new approach still in the feasibility stage, which could be used to monitor the inner walls of chemical process piping. A fiber optic sensor to measure lignin concentration in wood pulp has been demonstrated[17] and was described at this symposium last year. Development of a commercial system for the pulp and paper industry is now underway using this sensing approach.

3.4 Manufacturing

Optical and fiber optic sensors pervade manufacturing operations. Fiber optic temperature, pressure, proximity, and pushbutton shutter and interrupter switches are used in many production line applications. High resolution optical imaging systems now provide not only monitoring functions, but on-line control and gaging of dimensions as well.

Figure 10 shows the output of a noncontact fiber optic weld bead temperature monitor used to measure the temperature profile across a seam weld made between the edges of a continuously moving part. The convergence of the edges just prior to welding is monitored with a line scan camera assembly shown in Figure 11. Both sensors are used simultaneously to control the high frequency electric resistance welding (HFERW) process. The weld bead temperature profile is determined using two color analysis of the Planck blackbody radiation profile[18]. The gap between the edges in the convergence region is monitored with the line scan camera, as explained in the following paragraphs.

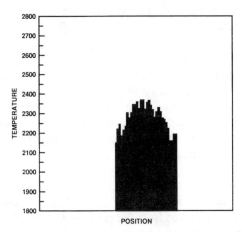

Fig. 10. Weld bead temperature monitor output profile of temperature versus position.

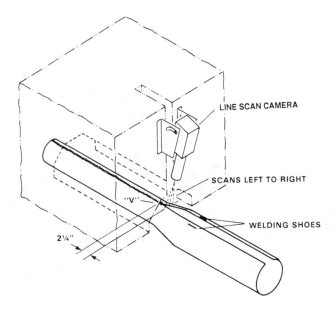

Fig. 11. Line scan camera mounted on mill for tube production.

In the high frequency electric resistance welding process (HFERW) for mechanical tubing production, plates of flat, strip steel are bent and formed into a smooth cylindrical shape. Before the edges of the strip are pressed together, welding shoes contact the strip near each edge. A large electric current from the HFERW power supply passes into the steel to heat the edges close to the melting temperature. The edges are then forged (pressed) together between weld pressure rolls in a special mill.

The diagram in Figure 11 shows a tube in the intermediate and final stages of formation. Just prior to forging, the edges of the cylindrical plate converge after the welding shoes to form a "V". At the apex of the V (fusion point), forging begins. It is important that the V pattern, as determined by the convergence angle of the plate edges, be maintained constant. Pulsations of the strip caused by slippage during forming can lead to fluctuations in the convergence angle and instability in the position of the fusion point. This instability may result in weld nonuniformities and potential weld defects.

A practical method to monitor the V-convergence uses a line scan camera containing a 1024-element photodiode array. The set-up is shown in Figure 11. The hot, glowing edges of the strip following the welding shoes provide ample light intensity so that no external light source is required in this application. The camera is mounted to the structural frame of the mill machine and is aimed at a known position between the convergence point and welding shoes. An image of the gap across the V (from edge to edge) is focused on the photodiode array located at the camera image plane.

Typical output signals from the camera processing electronics are shown in Figure 12. Regions of high light intensity are received from the hot metal edges of the strip. Regions of low light intensity are received from the gap between the edges. The width of the gap is computed from the dark segment between the hot edges.

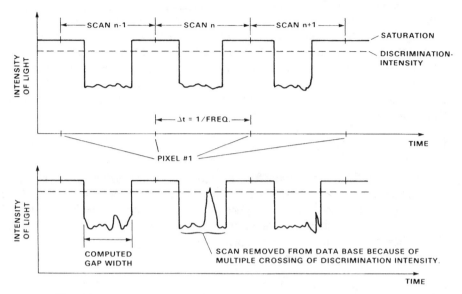

Fig. 12. Typical output signals versus time from line scan camera.

Mean values and standard deviation of gap width may be computed and displayed along with the light intensity versus time shown in Figure 12. The actual width is determined by counting the number of pixels across the gap and multiplying the result by a scale factor. Resolution in the width of .0015 inch can be achieved at steel strip line speeds of 30 inches/second.

New opportunities for fiber sensors in manufacturing processes and structural monitoring may result from the demonstrated ability to embed fibers in metals[19,20]. This capability offers the possibility of distributed stress and strain monitoring in metal structures similar to the work being done in composites. Other potential applications include monitoring the heat treatment of complex metallic shapes.

3.5   Power generation

Many demonstrations of fiber optic sensors are underway in the utility industry. Methods have been described[21] to measure flame temperature distribution, using Planck blackbody radiation from the flame, and two wavelength analysis of this radiation. A commercial product called a Video Flame Temperature Monitor is based on this early work. This product is about to be introduced by Diamond Power Specialty Company, Lancaster, OH.

Fourier transform infrared (FTIR) spectroscopy methods are being applied to the analysis of the many constituents in the flue gas of fossil fired boilers. Several organizations (one of which is Advanced Fuel Research, East Hartford, CT) are working to develop commercial FTIR systems for this application.

A low-cost fiber optic current sensor is being developed jointly by Centerior in Energy and Edjewise Sensor Products[22]. Light is transmitted between two butt-coupled multimode fibers. The light intensity is modulated by current-induced magnetic field changes. Changes in magnetic field misalign the coupled fibers, one of which is attached to a fixed anvil and the other to a ferromagnetic reed.

Pressure transducer development for power generation applications is continuing. A prototype fiber optic pressure transducer to measure primary water pressure in a nuclear steam generator is being built by Babcock & Wilcox. Metricor pressure sensors are presently under evaluation by several utilities for special applications. Luxtron, Mountain View, CA, and Paroscientific, Redmond, WA, are developing high precision fiber optic pressure transducers that measure the pressure dependent vibration frequency of a quartz crystal element.

3.6 Undersea and offshore applications

Interest exists in fiber optic sensors for undersea and offshore drilling platform use. One application is structural stress and strain monitoring of the undersea legs which suspend offshore platforms. Other potential applications for fiber optic sensors include downhole monitoring of pressure in the well, and inertial sensors to guide the well drilling apparatus. Most of these applications require wet-mateable fiber optic connectors. One such connector is made by Aurora Optics, Blue Bell, PA.

3.7 Medical applications

Medical applications of fiber optic sensors are beyond the scope of this paper. However, these applications are numerous and could be covered in a review paper of their own. Medical sensors are needed for both diagnostics and therapeutics and these applications are potentially very high volume. Proprietary activity in fiber optic sensors is underway in many medical service companies. Fiber sensors have been demonstrated for measurement of the partial pressure of the blood gases CO_2 and O_2. In-catheter measurement of pressure with disposable probes is commercially available from FiberOptic Sensor Technologies, Ann Arbor, MI.

3.8 Robotics

Robotics is a significant application area for fiber optic sensors which is addressed only briefly. Position measurement and control of articulated joints is one area where fiber sensors are being used. A robot hand has been developed[23] whose movement mimics that of a human hand inside a special glove. The position of the digits of the human hand is determined with fiber sensors attached to and embedded in the glove. The outputs of these fiber sensors are used to drive and control the digits of the robot hand.

Experiments are underway to evaluate on-board fiber sensors and fiber data links to remotely operated vehicles (ROVs). Concepts for remotely powered sensors on ROVs are based on work done by Little[24].

3.9 SSC and Maglev

Both the superconducting super collider (SSC) and magnetic levitation (Maglev) train are presently in the design phase, and fiber optic sensors are under consideration for applications to both. For the SSC, one need is to measure long-term changes in the position of the magnet field axis with respect to the axis of the proton beam tube. For the Maglev train, continuous monitoring of the structural integrity (stress and strain) of the rails is necessary. Both point and distributed fiber optic intensity and interferometric sensors are being evaluated for both applications.

4. STANDARDS AND SPECIFICATIONS

Progress is being made in fiber optic sensor standards activities within EIA, IEEE, and SAE. The Society of Automotive Engineers AS3-B Subcommittee has so far issued two aerospace resource documents SAE ARD50024 "Fiber Optic Coupled Sensors for Aerospace Applications" and SAE ARD50020 "Fiber Optic Interconnection Hardware for Aerospace Applications". The Electronic Industries Association has established standard test procedures for sensors and other components as described in TIA-455-A.

Several fiber optic sensor users and potential users are now writing fiber-optic-specific sensor specifications. Two potential users engaged in this activity are the Naval Ship Systems Engineering Station, Philadelphia, PA, and the Electric Power Research Institute (EPRI), Palo Alto, CA. In order to facilitate interaction with EPRI during this activity, and to help bring optical sensors to the market place, an Optical Sensing Manufacturers and Utilities Group (OPSM/UG) has been formed.

5. IMPORTANT ISSUES

In this section, several issues on fiber optic sensors are discussed. Some of these issues have been addressed previously[25], and several new issues are discussed as well.

Unanswered questions still exist on the reliability and maintainability of fiber optic sensors. The situation is not much different than when last examined four years ago. Sufficient quantities of identical sensors have not yet been produced to enable statistically significant failure analyses to be performed. Until large numbers of fiber optic sensors are used in service, a database of application specific performance cannot be generated.

In regard to support components such as connectors and couplers, much better information is available. User familiarity with these components is growing and applications are becoming more widespread. Multimode ST-style connectors and fused bi-directional couplers have excellent performance from the standpoint of cyclic repeatability and unit-to-unit repeatability. For example, the insertion loss of ST connectors with 100/140 fiber can be expected to vary less than 0.1 dB over 1000 connect/disconnect cycles[26]. Variation in insertion loss and coupling ratio from 100/140 fused couplers may be expected to vary less than 0.25 dB and ±5%, respectively, from unit to unit[27]. Such performance builds user confidence in fiber optics technology and this experience base carries over into more complex systems such as fiber sensors, which contain these components.

Information currently available on fiber optic multimode intensity sensor designs seems to indicate that for mass production units the per-unit cost will be low. In some cases (accelerometers and pressure transducers especially), the manufactured cost may be less than commercially available sensor units that use competitive transduction technologies such as piezoelectric, resistance, capacitance, or inductance. This opinion is based on the relative simplicity of signal processors for fiber optic intensity sensors, and the mechanical simplicity of the sensors themselves.

Furthermore, fiber optic sensors can be built to be smart. Self-referencing methods described earlier provide the means to null out intensity changes caused by connectors and link losses. Compensation for thermally induced output changes can be built into the sensor head. For example, with a differential pair of microbend sensors, both sensors can be arranged to detect environmental temperature changes with only one sensor arranged to detect changes in the displacement measurement of interest. The electronic signal processor which deciphers the optically encoded information from the two sensors can be located thousands of feet away from the sensors with no electric or metallic interconnections required.

Although fiber optic intensity sensor designs are the most commercially advanced at this time, improved performance from some interferometric sensors indicate that these may soon be ready for advanced design and prototyping. In-fiber Fabry-Perot sensors have been demonstrated for extended range temperature measurement[28], and these sensors embedded in composites have been used for stress and strain measurement, thermoset cure, and as acoustic emission monitors[29]. Further improvements in this technology are needed to simplify the fabrication of in-fiber Fabry-Perot sensors and to provide Fourier transform signal processors which can extract the sensor encoded phase information from the phase noise in the rest of the link[30].

6. CONCLUSIONS

Fiber optic sensors are making inroads into specialty application areas where the technology offers a unique advantage over conventional methods. Most fiber optic sensor equipment continues to be sold for test and evaluation purposes, and these installations are providing the baseline data from which specifications for future systems will be written.

Notable improvements have been made in the performance of many fiber optic sensors. Continued improvements will be made as the engineering of existing sensors into suitable packages evolves. This engineering process can tend to be tedious and expensive, and thus will proceed at its own pace based on customer demands and availability of funds.

Most importantly, competitive costs will result from the availability of better engineered units. Fiber sensors will continue to exploit the many advantages offered by fiber optics technology. This means that fiber sensor systems will be engineered with the following attributes:

- without the need for electrical power or local signal processors at the sensor
- with many sensors optically multiplexed to a common bus fiber or to several bus fibers on the same network to improve reliability
- with the optical demultiplexer, optical-to-electronic converter, and smart signal processor located at the end of the bus thousands of feet from the individual sensors

The improvements in fiber sensor technology to support this architecture are being made now and, as a result, prognostications for the future of fiber sensors are very positive.

7. ACKNOWLEDGMENTS

Some of the work described in this paper was supported by McDonnell Douglas Corporation (Contract Z0004) and General Electric Company (Contract MAO C-221). The author wishes to thank the following organizations for providing material for the presentation: Battelle Columbus Laboratories, Eldec Corporation, Electric Power Research Institute, FIMOD Corporation, and Luxtron Corporation.

REFERENCES

1. J.W. Berthold, D. Varshneya, and W.L. Ghering, "Calibration of High Temperature Fiber Optic Microbend Pressure Transducer," *Proc. SPIE 718*, p. 153, Fiber/Lase '86, Cambridge, Massachusetts, 1986.

2. J.W. Berthold, W.L. Ghering, and D. Varshneya, "Design and Characterization of a High Temperature Fiber Optic Pressure Transducer," *J. Lightwave Tech. LT-5*, p. 870, July, 1987.

3. S.E. Reed and J.W. Berthold, "Absolute Fiber Optic Pressure Transducer for Aircraft Air Data Measurement," IEEE NAECON, Dayton, Ohio, 1991.

4. Optech Microbend Accelerometer, 1989 Photonics Circle of Excellence Award, Photonics Spectra, April 1989.

5. D.R. Miers, D. Raj, and J.W. Berthold, "Design and Characterization of Fiber Optic Accelerometers," *Proc. SPIE 838*, p. 314, OE/Fibers '87, San Diego, California, 1987.

6. A.A. Boiarski and N.E. Nilsson, "Initial Field Application of a Distributed Fiber Optic Temperature Sensor Using Time Domain Reflectometry," Proc. of EPRI Workshop on Optical Sensing in Utility Applications, San Francisco, California, 1991.

7. K.A. Wickersheim, "Recent Developments in Fiber Optic Monitoring of Winding Temperatures in Power Transformers," Proc. of EPRI Workshop on Optical Sensing in Utility Applications, San Francisco, California, 1991.

8. N.E. Lewis and M.B. Miller, "Wavelength Division Multiplexed Fiber Optic Sensors for Aircraft Applications," *Proc. SPIE 989*, OE/Fibers '88, Boston, Massachusetts, 1988.

9. D. Varshneya and W.L. Glomb, "Applications of TDM and WDM to Digital Optical Code Plates," *Proc. SPIE 838*, p. 210, OE/Fibers '87, San Diego, California, 1987.

10. M.T. Woldarczyk, "Environmentally Insensitive Commercial Pressure Sensor," *Proc. SPIE 1368*, OE/Fibers '90, San Jose, California, 1990.

11. T.A. Lindsay, et al., "Standard Fiber Optic Sensor Interface for Aerospace Applications: Time Domain Intensity Normalization," *Proc. SPIE 989*, OE/Fibers '88, Boston, Massachusetts, 1988. Also see U.S. Patent No. 4,681,395.

12. P. Daley, et al., "In-Situ Detection of Organic Contaminants in Soils and Groundwater," Proc. of U.S. DOE Information Exchange Meeting on Characterization, Sensors, and Monitoring Technologies, Dallas, Texas, 1992.

13. J. Haas and R. Gammage, "In-Situ Fiber Optic Monitors for Site Characterization," Proc. of U.S. DOE Information Exchange Meeting on Characterization, Sensors, and Monitoring Technologies, Dallas, Texas, 1992.

14. M. A. Druy, et al., "Fiber Optic Remote FTIR Spectroscopy," *Proc. SPIE 1584*, p. 48, OE/Fibers '91, Boston, Massachusetts, 1991.

15. G. Salamon, General Electric Company — Knolls Atomic Power Laboratory, Schenectady, New York, private communication to author.

16. S.E. Reed and J.W. Berthold, "Development of a Microbend Strain Gage," SEM Fall Conference on Experimental Mechanics, Keystone, Colorado, 1986.

17. L.A. Jeffers and M. L. Malito, "Continuous On-line Measurement of Lignin in Wood Pulp," *Proc. SPIE 1587*, OE/Fibers '91, Boston, Massachusetts, 1991.

18. L.A. Jeffers, "Microcomputer Based System for Remote Temperature and Control of an Induction Heated Process," ISA 31st International Instrumentation Symposium, San Diego, California, 1985.

19. C.E. Lee, et al., "Metal-Embedded Fiber Optic Fabry-Perot Sensors," *Opt. Lett. 16*, 1990 (1991).

20. J.J. Kidwell and J.W. Berthold, "Metal-Embedded Optical Fiber Pressure Sensor," Proc. SPIE 1367, p. 192, OE/Fibers '90, San Jose, CA, 1990.

21. E. Udd, in *Fiber Optic Sensors*, E. Udd, ed. (Wiley, New York, 1991), Chap. 13.

22. P. Shamray-Bertaud and E.F. Carome, "Lower Cost Fiber Optic Current Sensor for Electric Power Transmission and Distribution Systems," Proc. of EPRI Workshop on Optical Sensing in Utility Applications, San Francisco, California, 1991.

23. S.M. Wright, "Optical Attenuator Movement Detection System," U.S. Patent No. 4,972,074.

24. W.R. Little, "New Approach to Sensors for Shipboard Use", *Proc. SPIE 840*, O/E Fibers '87, San Diego, California, 1987.

25. J.W. Berthold, "Why Buy Fiber Optic Sensors?" Invited Paper, *Proc. SPIE 985*, p. 2, OE/ Fibers '88, Boston, MA, 1988.

26. Published data available from Molex® Fiber Optic Interconnect Technologies, Inc. and Optical Fiber Technologies, Inc. (OFTI).

27. Published data available from Gould, Inc., Fiber Optics Division.

28. C.E. Lee, et al., "Performance of a Fiber Optic Temperature Sensor From -200 to 1050°C," *Opt. Lett. 13*, 1038 (1988).

29. R.M. Measures and K. Liu, "Fiber Optic Sensors Focus on Smart Systems," *IEEE Circuits and Devices*, 8, 37 (1992).

30. D.A. Norton, "System for Absolute Measurements by Interferometric Sensors," *Proc. SPIE 1795*, OE/Fibers '92, Boston, Massachusetts, 1992.